he Codebreaker's Handbook

Brennan is a professional writer whose work has
d in more than fifty countries. He began a career in
sm at the age of eighteen and when he was twenty-four
the youngest newspaper editor in his native Ireland.

nid-twenties, he had published his first novel, an
l romance brought out by Doubleday in New York. At
y, he made the decision to devote his time to full-length
f fiction for both adults and children. Since then he has
ed more than one hundred books, many of them
onal best-sellers, for both adults and children.

Other books by Herbie Brennan
published by Faber & Faber

Space Quest
111 peculiar questions about the universe and beyond

The Ghosthunter's Handbook

The Aliens Handbook

The Spy's Handbook

Herbie Brennan's Forbidden Truths: Atlantis

The Codebreaker's Handbook

Herbie Brennan

Illustrated by The Maltings Partnership

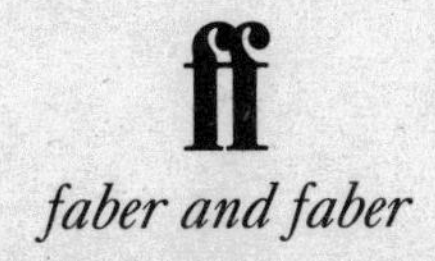

First published in 2006 by Faber and Faber Limited
3 Queen Square, London WC1N 3AU

Designed by Andy Summers, Planet Creative

Editorial management: Paula Borton

Illustrated by the Maltings Partnership

Printed in England by Bookmarque Ltd

A CIP record for this book is available from the British Library

ISBN 0-571-22461 X

2 4 6 8 10 9 7 5 3 1

Contents

This one is for JAB, with thanks for the help!

1 The Secret Message That Saved Greece

In 484 BC, the King of Persia was a hard man. His father, Darius, had been bad enough, but young Xerxes (pronounced *Serk-seas*) was a whole lot worse. A ruler with a very short fuse, he took no nonsense from anybody, especially any foreigner he looked on as his inferior ... and he looked on every foreigner as his inferior.

Before he died, the old King Darius had been having a bit of trouble with Egypt, which he'd been trying to sort out in a softly-softly way. When Xerxes came to the throne, he decided to take a tougher line. So he promptly raised an army and savaged the entire Nile delta. The Egyptians backed off and behaved themselves a whole lot better after that.

Then the city-state of Babylon started to play up. Xerxes sent his son-in-law to sort things out. In a vicious, violent campaign, every fortress in the city was razed to the ground, every temple pillaged. The Persians even went so far as to smash the statue of Marduk, the Babylonian's chief god – a hideous insult to any nation at the time.

Things quieted down quite a lot when that campaign finished. Xerxes stopped scowling so much and devoted himself to peaceful activities, like building a new capital for his empire. But his closest advisers wouldn't let him alone: they wanted more war.

His cousin Mardonius was the worst of them. He kept telling Xerxes he should avenge the defeat his father had suffered at the hands of the Greeks during the Battle of Marathon in 490 BC. Mardonius had a large following of exiled Greeks who, cross with their old government, added their voices to his demands.

But what really swung it was the homeland Greeks themselves. As word of Xerxes's great building project spread, celebratory gifts and tributes began to flow in from all over the Persian Empire and from neighbouring states who wanted to keep in the King's good books. But, as Mardonius was quick to point out, nothing at all appeared from Sparta or Athens.

It was all too much for the short-tempered Xerxes. "We shall extend the Empire of Persia such that its boundaries will be God's own sky," he told his Court, "so that the sun shall not look down upon any land beyond the boundaries of what is our own." Then he set out to make this proud prediction a reality.

For the next three years, Xerxes devoted himself entirely to preparations for an attack on the Greeks. Athens and Sparta, the two great Greek city-states, would be tough nuts to crack and he didn't

Xerxes

underestimate them for a second. He dispatched messengers to every governor who administered the far-flung outposts of his empire and demanded they send troops. He began a programme of shipbuilding and conscripted an enormous navy to ensure his army would have a secure supply-line. He embarked on a diplomatic offensive to rally support for his plans.

Eventually everything was ready. Xerxes cut a fine figure as he left Sardis at the head of a massive army. And what an army it was. Early historians insist it consisted of fully 5,000,000 men (although modern estimates are considerably smaller) supported by 800 ships. Among the officers was a man Xerxes believed to be his friend. His name was Demaratus.

Demaratus had been one-time King of Sparta, sharing the throne with Cleomenes I. There wasn't a lot of love lost between the two rulers. Demaratus blocked Cleomenes' plans to attack Athens and Cleomenes responded by (falsely) accusing Demaratus of being born illegitimate.

Since illegitimacy was a bar to kingship in those days[1], Demaratus was deposed and forced to flee to Persia. There Xerxes gave him a few small cities in northwestern Asia Minor to look after and Demaratus showed his gratitude by giving Xerxes help and advice when needed.

But while Demaratus marched with Xerxes at the start of the great invasion, he wasn't happy. Despite his exile, he still had feelings for his native Sparta and eventually decided he had to warn his fellow countrymen of the impending attack.

Xerxes and his men marched towards the Hellespont (a narrow strait linking the Aegean Sea with the Sea of Marmara and now called the Dardanelles) as Demaratus

1. And still is, in most parts of the world so far as I know.

pondered his dilemma. There was no way he could go on ahead without being missed, so he had to send a message of some sort. His problem was that the road was crawling with Persian guards and if they intercepted the communication, Sparta would still be in trouble and Demaratus would be minus his head.

So he did something that was very unusual in those days – he sent a secret message. This was not the first secret message in recorded history, but it came close. The very first was sent in the time of King Darius, Xerxes's father, by an unsavoury character called Histiaeus.

Histiaeus was trying to stir up trouble for Darius and wanted his son-in-law to start a revolt. He had, of course, exactly the same problem that would face Demaratus a couple of decades later – if the message was intercepted, he was dead. So he did what Demaratus did and conceived a cunning plan to conceal the missive.

I'll tell you what happened to both these secret messages in a minute. But before I do, why don't you try to figure out how they were sent? As a little help, I should mention there were no codes or ciphers involved and invisible ink hadn't been invented yet. To work out how Demaratus did it, you'll need to find out how ordinary, non-secret, letters were sent at that time[2]. But Histiaeus was just plain clever and used an ingenious ruse you might even be tempted to try today.

You'll find details of how both messages were sent and received in the box at the end of this chapter, but for now, let's find out how things turned out after they *were* sent. The first thing to say is that both messages got through safely.

In the case of a Histiaeus, he was successful in triggering a revolt, but then had the cheek to persuade

2. Hint: research it on the Internet.

Darius that he (Histiaeus) was the very man to put it down. While engaged in this complicated scheme, his loyalty came under suspicion and he fled to a new career as a pirate based in Byzantium. In the end he was captured and crucified at Sardis.

Demaratus was rather more successful. Xerxes placed two boat bridges across the Hellespont, but a storm blew up and destroyed them. In a fit of pique, he had the sea whipped as punishment for trying to thwart his plans[3]. This firm action did the trick and the sea remained calm as the bridges were rebuilt. It took fully seven days for his massive army to pass over them.

To avoid having to climb Mount Athos, he had a channel dug across the Isthmus of Actium and was soon engaged with the Greeks. At first the war went well. Xerxes won the Battle of Thermopylae in August and the following month occupied Attica and pillaged Athens. But despite appearances, his overall campaign was in trouble.

Forewarned by Demaratus's message, the Greeks hurried to arm themselves. For years, all profits from the state-owned silver mines were shared among the citizens. Now they were diverted into boat-building. As a result, the Greek navy had 200 warships nobody knew about. It proved to be a critical factor in deciding the outcome of the entire war.

The crunch came at the Bay of Salamis, just along the coast from Athens. At first the Persians were content to wait, knowing their fleet was vastly superior on the open sea. The Greeks, on the other hand, did their level best to entice the Persians into the bay where their own more manoeuvrable ships waited in ambush.

3. People did odd things like that in olden times. The mad Roman Emperor, Caligula, once went to war with Neptune, god of the sea, proclaimed himself the winner (since Neptune failed to turn up) and returned to Rome with chests full of seashells as booty.

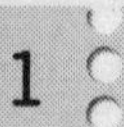

As it turned out, the gods favoured the Greeks. A change in wind direction blew the Persian fleet into the bay and engagement took place on Greek terms. The battle quickly turned into a terrible disaster for the Persians. Xerxes's fleet was under the command of Queen Artemisia of Halicarnassus whose own ship was quickly surrounded. She managed to break free and headed for the open sea, but rammed another Persian ship in the process. Both crews panicked and more Persian ships were rammed. The Greeks took advantage of the confusion to launch a full-scale attack.

By the end of the day, the Persian fleet was in tatters. Queen Artemisia advised Xerxes that without the navy

guaranteeing his supplies, he could not afford to continue with the war. Xerxes took the advice and withdrew. It proved the beginning of the end for his entire empire ... and all because of a single, secret message.

How the messages were sent

During the reign of Xerxes, all letters were written on wooden-backed wax tablets. A sharply pointed stylus was used to scratch the words into the wax which could be melted down and re-used once the message was read. The ancient historian Herodotus explains how Demaratus got his message past the guards:

Demaratus framed the following contrivance. He took a pair of tablets, and, clearing the wax away from them, wrote what the king was purposing to do upon the wood whereof the tablets were made; having done this, he spread the wax once more over the writing, and so sent it.

By these means, the guards placed to watch the roads, observing nothing but a blank tablet, were sure to give no trouble to the bearer.

When the tablet reached Lacedaemon, there was no one, I understand, who could find out the secret, till Gorgo, the daughter of Cleomenes and wife of Leonidas, discovered it, and told the others.

"If they would scrape the wax off the tablet," she said, "they would be sure to find the writing upon the wood."

The Lacedaemonians took her advice, found the writing, and read it; after which they sent it round to the other Greeks. Such then is the account which is given of this matter.

Histiaeus took an even simpler – and rather more ingenious – approach. He shaved the head of his messenger, wrote his letter on the man's bald pate, then waited for the hair to grow again before dispatching him. When he reached his destination, the man had a quick haircut, then bowed his head towards the recipient of the message.

2 The Odd Art of Invisibility

Steganography sounds like the name of a prehistoric reptile, but it isn't. It's the art of hiding secret messages. Every nation in the world has shown an interest in steganography at one time or another because every nation in the world has had secrets it wanted to keep. Some of the methods steganographers dreamed up were positively disgusting.

One was the use of pee. Not many people know this[4], but urine makes excellent invisible ink – and has the added benefit of always being on tap, so to speak. For this reason, it's long been popular with spies, who manufacture their own supplies on a daily basis.

To try it out – although I've a feeling I shouldn't be encouraging you to do this – simply pee in a cup and use the liquid to write your message on a plain piece of paper. You'll be able to see what you're writing, if only just, but once the ink dries, it disappears. (Usually. The actual chemical composition of your urine depends on what you've been eating and certain foods, not to mention some vitamin pills, can colour it enough to leave visible traces when the ink dries out. If this happens to you, try drinking lots of water – which is good for you anyway – until the flow starts to run reasonably clear. You might

4. Unless, of course, they've read my book *The Spy's Handbook*, Faber and Faber, London, 2003.

also try to avoid eating asparagus. It doesn't colour urine strongly, but it does make it smell so bad that anybody with a sharp nose may suspect a secret message on the paper.)

Once your secret message is in place, you can reveal it by using heat. Iron the paper, leave it on a radiator or whatever, and the writing reappears in a nice shade of brown.

There are a great many other things you can use to make invisible ink, most of them probably lying around your house somewhere. It's worth experimenting with almost any fruit juice, but the ones I can guarantee are lemon, lime and apple. Milk will do the trick as well, as will onion juice, a solution of sugar, diluted honey, diluted Coca-Cola (along with most other cola drinks) and vinegar.

Vinegar is a particularly interesting one. As with all the others mentioned, you can bring out vinegar writing with heat. But you can also bring it out by spraying it gently with red cabbage water.[5]

If you fancy yourself as a bit of a chemist, you can lash

5. For the technically-minded, vinegar changes the pH factor in red cabbage water, so it's not really the vinegar that appears, it's more that the cabbage water changes colour where the vinegar dried out.

up even more invisible inks by making solutions of baking soda (sodium bicarbonate) washing soda (sodium carbonate) Epsom salts (magnesium sulphate) or just plain old table salt. Other chemicals that work well are copper sulphate, ammonium chloride, iron sulphate, sodium nitrate and cobalt chloride – all in solution. Most dilute acids work too, but I really mean *dilute*, otherwise all you do is burn holes in the paper[6].

Everything mentioned so far becomes visible when you heat it, but you can get really fancy with some invisible inks by using a chemical reaction to make them appear. If you treat a lemon juice message with iodine solution, the ink turns white while the paper will turn light blue, which is a lot more impressive than the brown-on-white you get by heating it. Iodine solution will also bring up invisible ink made from household starch. (This gives you a very stylish blue on blue message: the ink turns dark blue, the paper light blue.)

Invisible Ink Recipes

To make ink:

Mix a tablespoon of baking soda and a tablespoon of water in a cup.

To make visible:

Paint with purple grape juice.

To make ink:

Squeeze a lemon or a lime into a clean container.

6. As with all chemicals, treat them with care, keep well away from your mouth and eyes, and wash hands thoroughly. Have an adult to hand.

2

To make visible:

Heat.

To make ink:

Mix two tablespoons of corn starch with four teaspoons of water, stir until smooth. Heat in a pan for several minutes.

To make visible:

Sponge some iodine solution over the paper.

And here's a really neat one. Stuff called *phenolphthalein* makes an invisible ink that turns a lovely fuchsia pink if you treat it with a solution of washing soda. If you can't find a sack of phenolphthalein lying about, try looking in the medicine cabinet for your old man's laxative tablets. Check the ingredients and with a bit of luck you'll find they contain phenolphthalein. If they do, crush half a dozen of the tablets using a spoon, mix with three or four tablespoons of water and you have your ink. (If the pills have a coloured coating, scrape it off before you crush them.)

To bring up the secret message, put some plain water on a dinner plate and use it to soak the paper. If you're using washing soda, mix four teaspoons in four tablespoons of hot water, let it cool and leave it aside until you're ready. When the message paper is sopping, you need to dribble the washing soda solution over it very carefully, allowing it to spread across the whole surface. Moments later you'll have your message back.

Invisible ink was used to considerable effect during the

Second World War, as shown by an old postcard that turned up in a sale a few years ago. The card was sent out of Nazi-occupied Poland and its simple one-paragraph message contained nothing to attract the attention of the German censors. But hidden on the card was another message, written in invisible ink. It described the horrors of a Nazi death camp and asked for supplies ... including more invisible ink.

And before you start thinking invisible ink is old hat for hiding messages, you should know that the CIA (Central Intelligence Agency) does not agree. (The CIA is America's secret service like Britain's MI5.) Back in 1998, lawyers for the James Madison Project (a non-profit organisation dedicated to reducing secrecy in public life) asked officials to release the oldest classified (i.e. secret) documents held in America's National Archives.

The officials came back with the names of six documents dating back to the First World War (1914–1918) and all dealing with invisible ink. (One was called *German Secret Ink Formula*, another *Pamphlet on Invisible Photography and Writing*.) But the Archive officials said they could not release the papers themselves, since this decision had to be made by the CIA.

Assuming there would be no problem with documents more than 80 years old, the lawyers applied to the CIA ... who refused point blank and continued to refuse in the face of court action. The Agency claimed that information wasn't necessarily useless just because it was old, and invisible inks were still in use by CIA officers and their sources.

3 Some Other (Really Neat) Ways to Hide a Message

Of course, not everybody relied on invisible ink to hide their secret messages. Take the Ancient Chinese, for example. China was the only country in the world able to produce silk. For centuries the way it was made remained a complete secret. But silk not only *was* a secret, it was also a means of *keeping* secrets.

Fine silk can be folded into an extremely small space, stage conjurers still use it to produce large streamers and banners apparently out of thin air. The Chinese would write a message on a piece of very finest silk, then roll it into a tiny ball. They then coated the ball with wax.

Thus prepared, the message was ready for dispatch. To keep it secret, the messenger swallowed it and hastened to his destination before he had to go to the loo. Once there, he squatted and strained until out popped the message. When it was washed off and the wax seal broken, it was no worse for its internal journey.

A much less smelly method of hiding a short secret message is to write it on the inside of an egg. The technique dates back to 15th-century Italy where it was discovered by a scientist named Giovanni Porta. First boil an egg until it's hard – about eight minutes should do the trick. Next, get yourself down to the chemist and buy yourself an ounce of alum. Mix the alum with a pint of vinegar and use it to write your message on the shell of the egg. (see footnote on page 17 concerning chemicals.)

Now here's the really interesting part. The alum/vinegar mixture won't show up on the outside: as soon as it dries all trace of your message disappears. So you can pop your doctored egg in a box with five others and send it off to a friend. If anybody opens the box, there's absolutely nothing to advertise that one of the eggs has been tampered with – even the fact it's hard boiled doesn't show. But when your friend shells the egg, there's your message in plain sight on the hard-boiled white.

This sort of nonsense is fine for short messages, but the time will come when you want to hide a long one. This happened a good many years ago to a distinguished Australian poet who was in the habit of contributing verse to a specialist magazine called *The Bulletin*. Her relationship with the magazine remained good for a time, but she eventually fell out with her editor about money and stormed off, vowing she would never write for the publication again unless she received the fee she demanded.

The editor dug in his heels and refused, predicting to his colleagues that the poet would back down. It turned out he was right. After a couple of months the woman submitted a lengthy two-part poem in blank verse. The first part was entitled *Abelard to Eloise* and the second *Eloise to Abelard.*

The editor smugly published the work and sent the poet her standard fee. Days after the magazine appeared on the news stands, somebody pointed out that if you read downwards the first letter of each line of verse, it spelled out a secret message. The message went something along the lines of: **SO LONG BULLETIN BOG OFF EDITOR.**

Messages like this are a bit easy to spot (unless you're a *Bulletin* editor) but you can hide one more effectively – and still in plain sight – by incorporating it into a crossword puzzle.

Even if you're not a crossword fan, you'll know this popular art form challenges readers to work out words from clever clues. If you create a crossword that contains every word of your secret message[7], then you can, at the same time, conceal and reveal it by sending your target nothing more than a handful of numbers. Here's how it works.

Suppose your secret message is **WATER THE TRIFFIDS AT ONCE**. Your crossword clues might read something like:

Hijklmno. (5.)

What were they in Wyndham's day? (3, 8.)

If not sooner, he ordered. (2, 4.)[8]

7. I know that isn't easy, but you have to make some effort in the secret message business.
8. H to O in the alphabet = H2O the chemical formula for water; the British author John Wyndham wrote the sci-fi classic *The Day of the Triffids* about murderous walking plants; 'At once if not sooner' is the sort of stupid thing adults say to you when they want you to do something right away.

Insert these clues anywhere you like in your developing crossword and fill up the rest with any old rubbish that occurs to you. Assume for a minute your secret message clues just happen to be 4 down for *What were they in Wyndham's day*, 11 across for *Hijklmno* and 22 across for *If not sooner, he ordered.* You can safely publish your crossword in *The Times, The Beano* or wherever. Even if readers crack your individual clues, there's nothing to suggest they go together. But if you send your target a note that simply says '22a, 4d, 11a' he'll not only know where to look for your message, but also the right order to place the words when he works them out.

The triffid business is a very simple message, but you can easily see that by creating a large crossword, quite complicated messages could be sent. Which is exactly what the British Secret Service thought was happening during World War Two when codenames for secret Allied military operations appeared in *The Times* crossword. They interviewed the compiler who turned out not to be the spy they expected. He'd picked a series of codewords by sheer coincidence.

All the same, that's the trouble with using a crossword. Any sensitive material in your message is likely to be spotted and before too long, the codebreakers will know exactly what you're up to. For that reason, steganographers often disguise secret messages as something else entirely, like a magazine article, a letter home or a shopping list.

The place where you hide the message is known as your *covertext*. You start with the perfectly innocent article, letter or list, then modified (changed) in some way to carry the hidden message. Here's a bit of modified covertext (known as *stegotext*) for you to examine:

It's dark. Pitch dark. You're seated with a group of people around a heavy wooden table. The room is so quiet you can hear your colleagues breathing. No-one speaks. No-one moves. You're quite sure of this because you're holding the hands of the people on either side of you and they're doing the same with the people on either side of them...and so on all the way round the table. You feel as if you've been sitting in the dark forever.

A cold breeze blows across the back of your neck as if somebody had opened a window or a door. But still nobody has moved. Then, without warning, a hand touches your face. You jerk back in alarm. "What's the matter?" asks the girl sitting next to you.

Before you can answer, there's a loud cracking sound, like a pistol shot, from the middle of the table. "What was that?" the girl asks in sudden alarm.

Another crack, louder this time. Then the table trembles ... shakes... shifts...jerks. Out of the darkness someone moans as the table begins to levitate...[9]

9. If you're desperate to read more, this covertext comes from my fascinating book *The Ghosthunter's Handbook*, published by Faber & Faber. Rush out and buy a copy now.

Can you spot the secret message? It's in there all right – it says *Buy Herbie's books* – but you might have trouble discovering exactly where. So here's the stegotext again with the modifications made a bit more obvious:

> It's dark. Pitch dark. You're seated with a group of people around a heavy wooden table. The room is so quiet you can hear your colleagues **b**reathing. No-one speaks. No-one moves. You're q**u**ite sure of this because **y**ou're **h**olding th**e** hands of the people on eithe**r** side of you and they're doing the same with the people on either side of them…and so on all the way round the table. You feel as if you've **b**een sitting **i**n th**e** dark forever.
>
> A cold breeze blow**s** across the **b**ack of y**o**ur neck as if someb**o**dy had opened a window or a door. But still nobody has moved. Then, without warning, a hand touches your face. You jer**k** back in alarm. "What'**s** the matter?" asks the girl sitting next to you.
>
> Before you can answer, there's a loud cracking sound, like a pistol shot, from the middle of the table. "What was that?" the girl asks in sudden alarm.
>
> Another crack, louder this time. Then the table trembles … shakes… shifts…jerks. Out of the darkness someone moans as the table begins to levitate…

Now each letter of the message is highlighted in bold typeface. But how was the first example of this stegotext modified? If you still can't spot it, don't worry – it was cunningly done. What I actually did was increase each letter of the secret message by a single point size. Unlike a bold typeface, it's not something you'd normally notice, but if you look very carefully, you'll see it's definitely there.

Once computers came into general use, steganographers had a fabulous new concealment tool, particularly since the advent of the Internet which, believe me, is now crawling with secret messages. Many of them are hidden not in text, but in pictures, since visual images and words are encoded by computers in exactly the same way – an arrangement of bits and bytes.

The larger the amount of data in your covertext, the easier it is to slip in a hidden message; and the longer the hidden message can be. Photographs, as you probably know if you run your own website, contain a heck of a lot of data. This makes them ideal for conversion into stegotext, even though they aren't actually text at all to begin with.

This is a little bit technical if you're not into computers, but here's how it's done. A simple bitmap (RGB) picture will have 8 bits representing the three primary colours red, green and blue. That gives you 2^8 (2 to the power of 8) different values for each colour. That's a *lot* of different values and the chances of the human eye telling the difference between a shade of red represented by 11111111 and a red coded as 11111110 are very slim indeed.

That means you can safely use the least significant *bit* of a colour coding for something else. Now comes the cool part. Computers code text as ASCII (American Standard Code for Information Interchange or some such nonsense) with three *bits* making up each letter of the alphabet. So by stealing a *bit* each from the red, blue and green colour coding of a picture, you can hide a letter of the alphabet in just three pixels. Given that most Internet pictures are composed of *thousands* of pixels, you can see there's lots of room in there for your message.

Even more sneakily, you can disguise your message further by making it look like the random pixels you get scattered through an image when you compress it into a jpeg, the most popular format for web pictures. And since computers treat digital audio the same way as digital pictures – all just *bits* and *bytes* remember – you can hide an ASCII message in a sound file as well, cunningly disguised as background noise.

And if you get tired of hiding text messages, you can even hide an entire picture. You'll need a good graphics application to do it, but if you remove all but the last two *bits* of the three RBG colour components of a photograph, the whole thing will appear to turn black.

But it isn't black really. All you need to do is boost the brightness 85 times and the image will reappear.

4 Julius Caesar, Urpdq Jhhchu

The only real drawback to steganography is that however cunningly you hide your message, you're always left with the uncomfortable feeling that if anybody does find it, your goose is cooked. For this reason, it wasn't long before people turned their attention towards cryptography.

Cryptography is a word that comes from the Greek *kryptos*, which means 'hidden.' But the idea isn't to hide your message, as in steganography, it's to hide the *meaning* of your message. In other words, you scramble it like an egg.

(You've probably already worked out that there's nothing to stop you scrambling your message, using a cryptography technique, *then* hiding it using one or other of the steganography methods you've been reading about. This gives you a *double* level of security – the message may never be found, but if it is, it can't be read – and is exactly what master spies do as a matter of routine.)

Cryptography not only comes from the Greek, but seems to have been started in Greece as well – or at least in Sparta, which was part of what we'd call Greece today. Back in the 5th century BC, some Spartan warrior with more brains than muscles developed what's now known as the Spartan scytale method of encryption.

A scytale is a plain wooden rod. To use it for encryption, a Spartan would take a strip of leather and wind it around the rod. With the leather in place, he then wrote his secret message along the length of the scytale. At this stage anybody could read it, but when he unwound the leather, all he had was a strip decorated with a jumbled string of meaningless letters. He then wrapped the leather around the waist of his tunic as a belt, with the letters concealed on the inside. (Steganography!!)

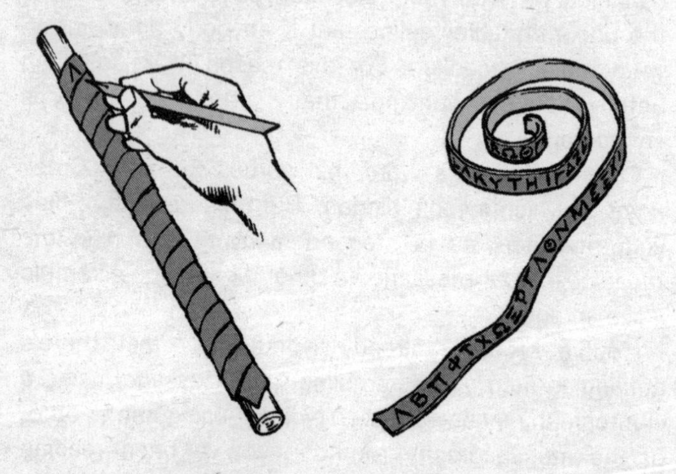

This method was put to good use in 404 BC when no fewer than five messengers set out from Persia (modern Iran) with urgent messages for the Spartan king, Lysander. Four of them failed to survive the journey, but the fifth got through. He stripped off his belt and handed it to his king, who wrapped it round a scytale *of exactly the same diameter as the one used to write the message.* At once the individual letters fell into place and the message became clear. Pharnabazus of Persia was planning to attack Sparta. Forewarned, Lysander made some military preparations and was able to turn the attack when it came.

Lysander wasn't the only ancient general to make use of secret messages. More than 300 years later, the great Roman leader, Julius Caesar, had devised so many ways of scrambling his communications that much historical writing was devoted to his ciphers.

The very first one he used – at least as far as surviving records show – was during his long wars in Gaul (now France). His friend and ally Cicero was under siege and on the point of surrender. Caesar wanted to tell him help was on the way, but obviously couldn't risk the enemy finding out about his planned troop movements. He hit on a simple, but highly ingenious way of encrypting the message, which I'll tell you about in a moment. Then he handed the letter to a messenger and told him to take it to Cicero. With foresight born of long experience, he instructed the messenger that if he couldn't get through the besieging forces, he was to tie the letter to a spear and throw it over the barricade.

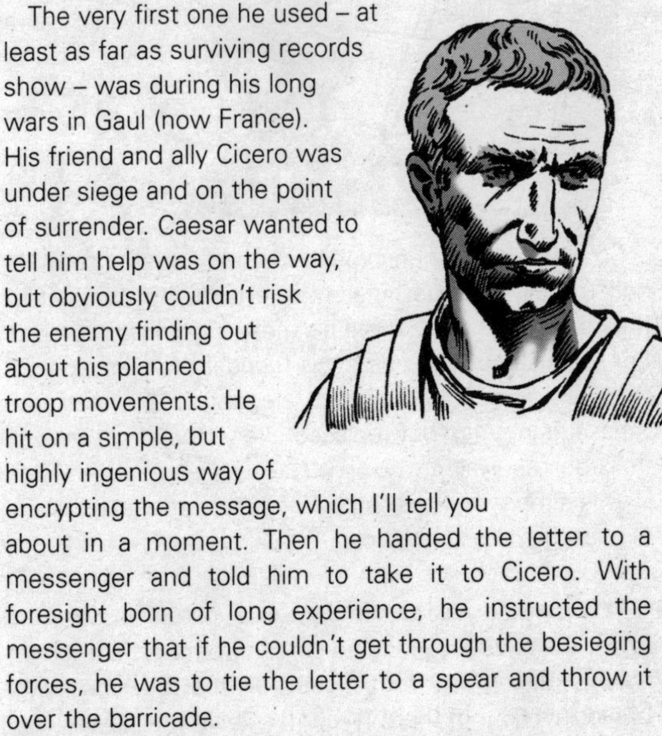

As it happened, the messenger did indeed have problems getting through and followed Caesar's instructions to the letter. But the spear stuck high up in a tower and nobody in Cicero's camp noticed it for three whole days. Eventually a soldier spotted it, found the message and took it to his commander. Although enciphered, Cicero understood it at once and read it to his troops as an encouragement to hang on just a little longer.

So what was the encryption method Caesar used? He had to find something that would be absolutely meaningless to the Gauls if his messenger was captured, but still instantly obvious to his friend if the message got through. His solution was elegant. He wrote the message in Latin, but replaced every Roman letter with its Greek equivalent.

This simple method can be incredibly effective. Take a look at the message below:

Looks like Ancient Egyptian, doesn't it? It's actually Shakespeare: "To be or not to be, that is the question." (I thought you could do with a bit of culture.) But unless you happen to read hieroglyphs you'd have a hard time working it out. I did just what Caesar did in his message to Cicero: substituted the alphabet of one language for another. But where he substituted Greek for Latin, I substituted hieroglyphs for English.

If you can get your hands on a personal computer, you can generate secret messages like this very easily. You probably aren't equipped with a hieroglyphic font, but if you look carefully through the fonts installed, you might well come across one called *Symbol, Dingbats* or something peculiar like that. Most computers carry special fonts designed to help mathematicians, graphic artists or business people. The expectation is that you'd use just one or two characters from these fonts, perhaps to decorate a graph or something like that. But you don't have to use them that way.

Try this experiment. Open up your word processor, check the font list and select one of the symbol fonts. Now type this message, ready to send to a friend:

I'M READING A FASCINATING BOOK AT THE MOMENT.

You'll find it won't come out looking anything like the way it looks above. Instead you'll get something like:

✩→✭ ✲✥✡✤✩✬✧ ✡ ✦✡✳✢✩✬✡✴✩✬✧ ✢✮✯✫ ✡✴ ✴✭✥
✭✮✭✥✬✴✎

(That's it in Zapf Dingbats)

or

✋⌫💣 ☼☜✌👎✋☠☝ ✌ ☞✌💧👍✋☠✌❄✋☠☝ 👌⚐⚐😐
✌❄ ❄☟☜ 💣⚐💣☜☠❄📪

(Wingdings)

or

ι∋μ ρεαδινγ α φασχινατινγ βοοκ ατ τηε μομεντ.

(Symbol)

You'll notice that *Symbol*, which at first glance seems as Greek as Caesar's message to Cicero, doesn't encrypt the entire message effectively. (Just look at the word 'book'.) The effect is even more pronounced if you type your message entirely in capitals, as I did with the first two examples. Then it comes out as:

ΙΠΜ ΡΕΑΔΙΝΓ Α ΦΑΣΧΙΝΑΤΙΝΓ ΒΟΟΚ ΑΤ ΤΗΕ ΜΟΜΕΝΤ.

Nearly half your message isn't encrypted at all.

Caesar took his chances with a simple Greek-for-Latin letter substitution in his message to Cicero. The Gauls might possibly read a bit of Latin since they'd been fighting the Romans for so long, but it was a rare one who'd know any Greek. The sophisticated Cicero, on the other hand, would have realised at once what was going on. But clearly you couldn't pull that stunt too often. Where the secret message had to pass through multilingual territory, some other form of encryption would be needed.

Later on in his military career, Caesar put his mind to the problem and came up with a method of encryption still used by schoolboys to the present day[10]. He invented what's known as the *substitution cipher*.

10. Although I'd bet good money very few of them know where it came from.

5 Sex and the Single Cipher

It was the Roman writer Suetonius who blew the whistle on Caesar's cipher but not until more than a century after the great Roman general was dead. Suetonius revealed that Caesar had hit on the idea of encrypting his messages by replacing each letter of the text with the letter that occurs three places along in the alphabet. Here's how it works, in Latin and in English:

A	B	C	D	E	F	G	H	I	J	K	L	M	N	O	P	Q	R	S	T	U	V	W	X	Y	Z
D	E	F	G	H	I	J	K	L	M	N	O	P	Q	R	S	T	U	V	W	X	Y	Z	A	B	C

In the first line, you have the ordinary alphabet. In the second, you have the same alphabet *shifted* three places along. I've used capital letters in both lines just to make your life a little easier, but I should mention that in the murky world of cryptography, plaintext is generally written in lower case and caps are reserved for cipher.

Since we've had enough military examples already in this book, let's suppose Caesar wanted to write an encrypted letter, beginning with the words *I love you*.

The plaintext (in Latin, of course) is *Ego te amo*[11]. Find

11. Please don't write in telling me the *Ego* is superfluous. I'm trying to show you how to encrypt a message, not write good Latin.

each letter of this message in the top line of Caesar's cipher printed here, then look directly below it for the letter you should substitute. E becomes H, G becomes J, O becomes R and so on. Eventually you'll encrypt the entire message, like this:

Plaintext = e g o t e a m o

Ciphertext = H J R W H D P R

While Suetonius tells us Caesar preferred a shift-3 cipher, you can see at once other shifts would work as well. With a shift-1 cipher, A would equal B, B would be replaced by C and so on. In a shift-4, A = E, B = F, C = G etc. But to be honest, the level of security you get from using a shifted alphabet isn't very high. Once somebody suspects you're shifting, he only has to check 25 possibilities to crack your cipher[12]. And that's a worst-case scenario. On average, he'll hit on the right shift every 12 or 13 tries.

Long after Caesar was murdered on the steps of the Forum, cryptographers worked hard to find a more secure approach. It turned out one was already in existence, but hidden away in some very obscure Hindu texts dating all the way back to the 4th century BC. About 300 years after Caesar's death, a Brahmin sage (wise man) named Vatsyayana drew on these texts to produce a book that's still read secretly under the bedclothes by teenagers to this day. He called it the *Kama Sutra*.

Sutra means 'book' and kama is a Sanskrit word for physical pleasure. When you discover Kama is also the God of Love in Indian mythology, you can guess the sort of physical pleasure we're talking about isn't confined to

12. Since there are 26 letters in the alphabet, there are only 25 ways you can shift it.

scratching your backside. For most people, the main interest lies in Vatsyayana's detailed descriptions of different sexual positions, but the book goes far beyond that. It contains instructions on many subjects including cooking, massage, perfume manufacture, book-binding, carpentry and enlargement of the human penis[13].

Somewhere along the line, Vatsyayana advises that women should make a study of *mlecchita-vikalpa*, the art of secret writing. He figured it would help them keep any notes about their love affairs strictly private and also let them send sweet little coded messages to attractive men without their wives finding out.

Rather than a shifted alphabet, Vatsyayana recommended a different approach which involved the random pairing of alphabet letters, like this:

A < >N

B < > Z

C < > P

D < > U

E < > O

F < > W

G < > R

H < > X

I < > Q

13. One method advocated is beating it with stinging nettles, something I've never been hugely tempted to try.

J < > Y

K < > T

L < > S

M < > V

The point here is that the pairings are purely random. I might just as easily have decided A < > X or B < > T. But once you've decided on your pairings, you can produce the *mlecchita-vikalpa* secret writing of Vatyayana simply by substituting partnered letters in your plaintext to produce the finished ciphertext. Thus, a note to your loved one might read:

XEF NZEDK N LAER, ZNZJ?

But if you wanted her to decipher it before your patience ran out, she has to know the pairings of the letters you were using. To make life easier for her, you might supply a special decoder like this:

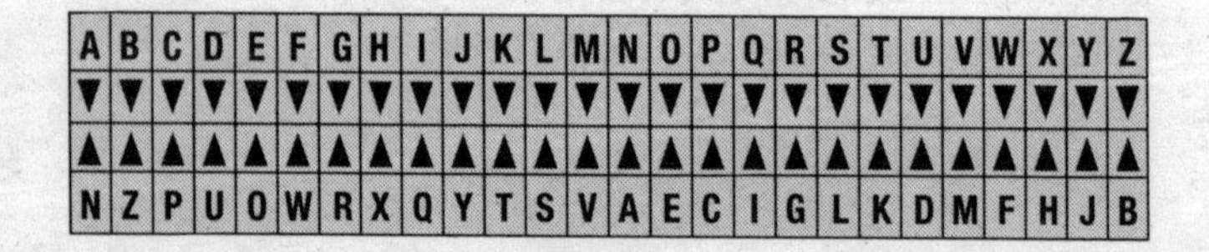

A	B	C	D	E	F	G	H	I	J	K	L	M	N	O	P	Q	R	S	T	U	V	W	X	Y	Z
▼	▼	▼	▼	▼	▼	▼	▼	▼	▼	▼	▼	▼	▼	▼	▼	▼	▼	▼	▼	▼	▼	▼	▼	▼	▼
▲	▲	▲	▲	▲	▲	▲	▲	▲	▲	▲	▲	▲	▲	▲	▲	▲	▲	▲	▲	▲	▲	▲	▲	▲	▲
N	Z	P	U	O	W	R	X	Q	Y	T	S	V	A	E	C	I	G	L	K	D	M	F	H	J	B

Which, of course, is where things start to get tricky. You'd obviously have to send her the decoder separately from your message and make sure nobody else saw it. She'd have to keep the decoder hidden as well. Furthermore, if you decided to change your pairings, you'd have to make up a whole new decoder and somehow get that to her (secretly) in time for her to read your next naughty message.

And frankly, even without a decoder Vatsyayana's secret writing isn't all that difficult to crack open provided you have enough time and a lot of patience. According to my calculations, there are 676 different ways you can pair off the letters of the alphabet[14]. That's a lot, but it's not an impossible lot. It might take you a few days or a few weeks to draw up decoders for every possible variation, but once you have them you could simply check the message against every one of them in sequence. Sooner or later, you'll find one that works and suddenly you're reading the juicy secrets of somebody's love life.

Vatsyayana's approach is approximately 27 times more secure than Caesar's shift-3 cipher, but it's still not good enough. The problem with both is regularity and rigidity. Caesar's approach allows you to shift a certain number of places, but once you shift, every letter falls in place and stays there. Vatsyayana did better, but his system actually starts out with letters rigidly paired, so it becomes possible to figure out what's going on if you're prepared to put in the work.

14. My calculations may be wrong. I'm not very good at maths.

6 Masonic Secrets

A completely different approach to the problem was devised by the ancient and mystic order of Freemasons, a secret organisation dedicated to rolling up one of its trouser legs. Since the Masons had a lot of business they wanted to keep private, the question of encryption became extremely important. They didn't trust the simple Caesar shift and Vatsyayana's pairings would never have been secure enough even if they'd been known in the West at the time.

Rather than rely on an existing cipher, one ingenious member of the Order indulged the Masonic love of symbolism by creating a new one of his own. He began by drawing up an intriguing grid pattern to hold the English alphabet. The grid looked like this:

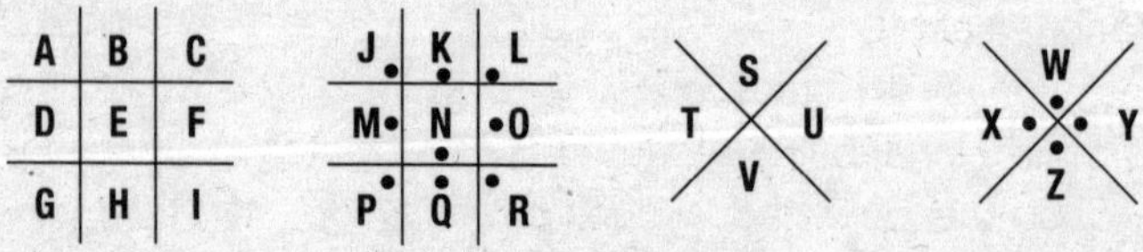

A B C
D E F
G H I

J• K• L•
M• N• •O
P• Q• R•

S
T U
V

W
X • • Y
Z

You might like to take five minutes to ponder how on earth you could turn this arrangement into a cipher, but if you can't come up with an answer, don't beat yourself up – it's far from obvious.

When the Masons wanted to send a message, what they did was to represent any given letter with the outline of its position in the grid. Thus:

A =

B =

C =

And so on.

If you fancy composing your own Masonic messages you can use the empty grid below as a guide. Just trace around the outline of where the letter was and don't forget the dots where they appear.

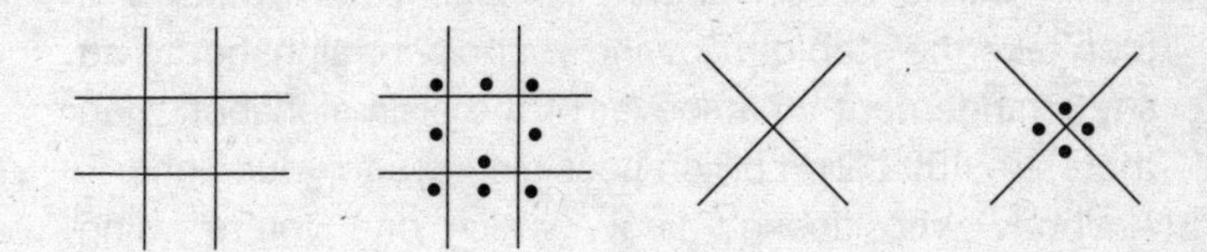

The result, you'll find, looks obscure, symbolic, mystical and extraordinarily impressive. Unfortunately it won't be terribly secure since Masonic writing is no longer secret. Once you know the grids, the cipher is easy-peasy to crack open. And even if you don't, the simple cipher-busting methods given later in this book will do the job without much difficulty.

Clearly, even the Masons failed to come up with an unbreakable cipher. Something more secure was needed.

7 Keyword Ciphers

As it happened, Vatsyayana came very close to solving the problem. Or rather putting Caesar and Vatsyayana together came very close to solving the problem. As long as you stick to a formal shift (Caesar) there are only 25 possible ways to break the cipher message. But if you start thinking about random pairings (Vatsyayana) and then take the step of allowing your cipher alphabet to be any arrangement whatsoever of the plain alphabet, then there are 400 billion billion possible keys to your cipher[15]. Anybody who doesn't know which one you're using would be long dead before he checked them all.

This isn't perfect security, as we'll see later, but it's still pretty impressive. To create your cipher, you simply match the plain alphabet with any random rearrangement of its letters, as in the example below:

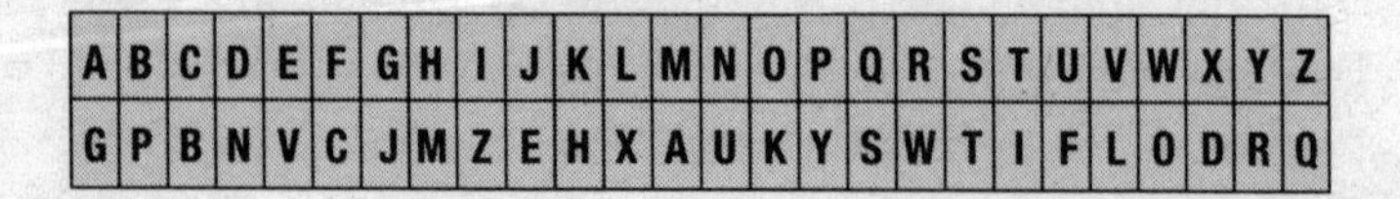

A	B	C	D	E	F	G	H	I	J	K	L	M	N	O	P	Q	R	S	T	U	V	W	X	Y	Z
G	P	B	N	V	C	J	M	Z	E	H	X	A	U	K	Y	S	W	T	I	F	L	O	D	R	Q

15. And that's the good old British billion, defined as a million million, not the miserly American billion which works out at a lowly thousand million.

(In the 10th century, Arab cryptologists managed to complicate matters by creating the world's first monoalphabetic substitution cipher in which some of the alphabet letters were replaced by symbols. Thus the A in plaintext might become ◆ in cipher, or an H could turn up as ✪. This was a bit like the Masonic approach, but didn't rely on any fixed grid and so added something to the overall level of security, but frankly not much. It did look rather pretty though.)

Once your target knows the order of the alphabet in your second line, deciphering the message is easy-peasy. But until she knows it, life is close to impossible. So the second line of the table above becomes the key to that particular cipher.

So how do you get the key to your target? You might just send it on a postcard, of course, but anybody coming across a sequence of letters like GPBNVCJMZ ... is likely to suspect you're up to something. Furthermore, if your friend happens to lose the postcard – and you know how careless your friends are – she'll have a very hard time remembering the random sequence of letters. And if she writes it down somewhere, she instantly doubles the chances of an enemy getting her hands on it.

You can begin to see the problem with a wholly randomised alphabet. The security is great, but the key leaves a lot to be desired. What you really need is something that gives you high security, but works off a key that's dead easy to remember and can be passed on without suspicion in a phone call or a letter. Fortunately that problem too was solved a long time ago when somebody came up with the idea of using a *keyword* or *keyphrase* to construct the cipher in the first place.

Keyword ciphers aren't *quite* as secure as a totally randomised alphabet, but they still generate so many possible keys as to be quite unbreakable by trial-and-error. And they have the huge benefit that passing on a keyword is both simple and suspicion free. Once she has the keyword, the target of your message can construct the entire cipher for herself in moments. You can even send 1,000[16] different messages using 1,000 wholly different ciphers, simply by passing on 1,000 keywords. Here's how it works.

First pick your keyword or keyphrase. It needs to be something easy to remember and so commonplace you could drop it into a letter or use it on the phone without attracting attention. Let's suppose you decide on the keyphrase *I love Brussels sprouts*[17].

Prepare your keyphrase for use by writing it down without any spaces and dropping any repeated letters. In the example above, this leaves you with **ILOVEBRUSPT**. Now start to build your cipher using the phrase as follows:

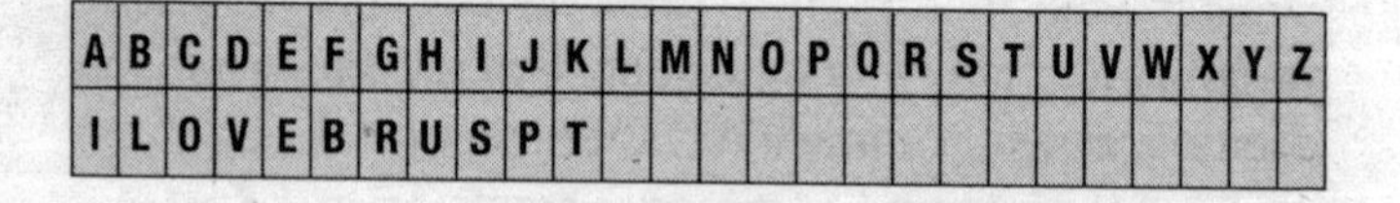

A	B	C	D	E	F	G	H	I	J	K	L	M	N	O	P	Q	R	S	T	U	V	W	X	Y	Z
I	L	O	V	E	B	R	U	S	P	T															

That's nearly half your cipher built already. For the remaining blanks, you simply start from where your keyphrase left off (in this case T) and fill in what's left of the alphabet in its usual order. The next letter in the alphabet after T is U, but we've already had that in our keyphrase, so you go on to V. Turns out we've had that as well, so we go on to W.

16. If that sounds a lot, remember you could simply use every word, in sequence, that appears on the first few pages of your favourite book. Let your target know the name of the book and you're in business.
17. Yes, I know. But it doesn't have to be true – it's just a keyphrase.

Since W doesn't appear in the keyphrase, that's what goes in after T. The letters after W are X, Y and Z, none of which appear in the keyphrase, so you put them in as well. Z is the last letter of the alphabet, so you go back to the beginning with A. Since A doesn't appear in your keyphrase, that goes in next. B comes after A, but B *does* appear in your keyphrase, so you go on to C. No sign of a C in 'I love Brussels sprouts' so in it goes. When you work your way right through, you end up with this cipher:

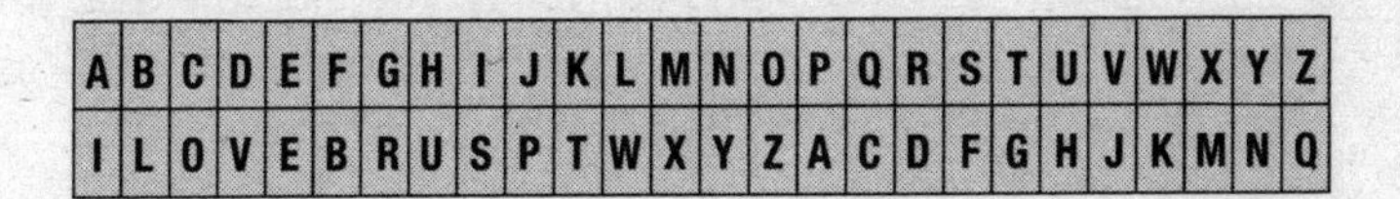

A	B	C	D	E	F	G	H	I	J	K	L	M	N	O	P	Q	R	S	T	U	V	W	X	Y	Z
I	L	O	V	E	B	R	U	S	P	T	W	X	Y	Z	A	C	D	F	G	H	J	K	M	N	Q

This approach dramatically increases the security of your cipher and might actually have changed the course of English history if it had been used by a group of conspirators towards the end of the 16th century. But it wasn't, so it didn't and the result, as you'll see in the next chapter, was that a queen lost her life.

8 The Cipher That Killed a Queen

8

Mary was kept under lock and key for 18 years before she decided to murder her cousin. The Mary in question was Mary Queen of Scots who had fled south from a military defeat in 1568, naively hoping Queen Elizabeth I would give her shelter. But cousin Elizabeth had her arrested instead. The charge – involvement in a murder – was trumped up. The real problem was Mary's religion – she was a Catholic and the vast majority of English Catholics considered her the rightful heir to the English throne.

Elizabeth didn't exactly throw her into the Tower, but she was closely confined in a series of castles and manors and, as the months turned into years, her situation became increasingly desperate. Then, early in the New Year of 1586, came a wholly unexpected development. A priest named Gilbert Gifford managed to smuggle some private correspondence in to Mary, who was confined at a place called Chartley Hall at the time. Gifford hit on the ingenious idea of wrapping the letters in a leather packet which he then hid inside the bung used to seal a barrel of beer. When the beer was delivered to Chartley, one of Mary's servants opened the bung and brought her the letters. (Steganography!) It quickly occurred to Mary she could smuggle letters *out* in the same way.

By this stage of British history, Catholics were having a hard time in England. The Mass was banned, members of the faith were heavily taxed and sheltering a priest was punished by tying you down, slitting open your stomach and forcing you to watch while your intestines were pulled out.

You can imagine this sort of treatment would breed resentment and nobody resented it more than a handsome young (Catholic) adventurer called Anthony Babington. One March evening in 1586, he convened a very secret meeting at the Plough Inn, near Temple Bar, and hatched a plot to free Queen Mary, assassinate Queen Elizabeth and start a rebellion that would put Mary on the throne.

Like any polite young man, he felt he needed Mary's permission to act in her name, but was faced with the problem of getting a message to her. He couldn't simply send a letter, which would be read by her guards, and he didn't know about Gifford's activities at this time. But Gifford knew about his and turned up on Babington's doorstep on July 6 offering to carry a message to the imprisoned Queen. Babington took him up on the offer and immediately composed a letter to Mary. Wisely, he decided to encrypt it.

The encryption he used was basically a substitution cipher, but unlike the systems we've examined so far, Babington's cipher substituted symbols for each letter of the alphabet in use in his day. You'll notice from the key below that the letters J, V and W are missing.

In an attempt to throw codebreakers even further off the scent, he used *nulls*, symbols like ff, ┌, ┘ and d that were peppered through the documents, but didn't actually mean anything at all: their sole purpose was to confuse. As an added layer of protection, a degree of code was also used.

Although the two words are often used carelessly to mean the same thing, there's a very real difference between a code and a cipher. A cipher substitutes something – a letter, a symbol, a sound or whatever – for *each letter of the alphabet* and the secret message is composed accordingly. A code substitutes something – usually a symbol – for every *word* of the secret message.

This means that the most famous code of all – Morse Code, shown in the table below – isn't a code at all, but a cipher.

Morse 'Code'

A •—	**B** —•••	**C** —•—•
D —••	**E** •	**F** ••—•
G ——•	**H** ••••	**I** ••
J •———	**K** —•—	**L** •—••
M ——	**N** —•	**O** ———
P •——•	**Q** ——•—	**R** •—•
S •••	**T** —	**U** ••—
V •••—	**W** •——	**X** —••—
Y —•——	**Z** ——••	

The code part of Babington's cipher covered some 35 words you might expect to find used fairly often in messages:

AND	2	RECEIVE	⫝
AS	n̲	SAY	ŋ
BEARER	T	SEND	ʃ
BUT	ʒ	SO	8
BY	ℴo	THAT	4
FOR	3	THE	8
FROM	⋇	THERE	ʊ
I	⊥	THIS	6
IF	4H	WAS	∂
IN	x	WHAT	m
IS	б	WHEN	‡

ME	ɱ	WHERE	ɤ
MEET	∃ℓ	WHICH	ɓ
MINE	SS	WITH	ɥ
MY	ɱ	YOU	⊣
NOT	x	YOUR	ʃ
OF	ɱ	YOUR NAME	ɜ
PRAY	⊢		

The strength of code is that it's near impossible to guess what word a particular symbol stands for if you don't have the key. The weakness is that you have to memorise a separate symbol for every word in the language, which isn't easy[18]. All the same, you could compose at least a few simple messages using only the code part of Babington's cipher. For example:

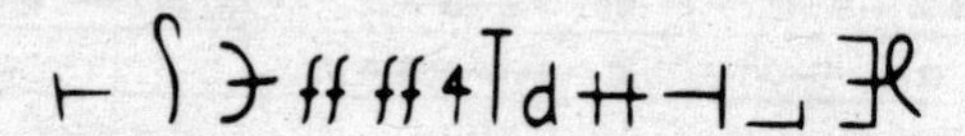

This decodes as *Pray send your name with bearer when you meet* (I dropped in a few nulls to confuse you.) The trouble is you can't send any really useful messages with such limited code, so the secret correspondence to the imprisoned Queen was a mix of code and cipher, mostly cipher, like this:

∇Ɛ∇‡ɑxΔΛ∇ƐnOϕ⧻ᚼ

8∞Oθθ|Δϸx‡n∇∇//

18. But not impossible. So far as I know, written Chinese works that way.

Babington's actual message was, of course, a lot less innocent than my example. He promised he'd work to rescue Queen Mary and 'dispatch the usurper' Elizabeth, claiming he had 100 followers and the backing of ten gentlemen, including six nobles, to help him in the venture.

True to his word, Gifford smuggled the message through the usual beer-bung route. On July 17, Mary sent back a reply, using the same cipher, in which she gave her blessing to the plot. She even stressed that she should be released before Elizabeth was assassinated, in case her jailers decided to kill her when they heard the news.

It looked as if Merrie England was all set for a palace revolution, but there was a fly in the ointment: Father Gifford was a double agent. Right from the start, he'd been in the pay of Sir Francis Walsingham, Queen Elizabeth's spymaster. Long before Babington's first letter got into the bung in the barrel, it was delivered to Walsingham's office.

Walsingham had no idea what was in it, of course. He couldn't tackle a complex ciper personally. But he knew a man who could. He handed the document to a short, slim, blond-haired scholar by the name of Thomas Phelippes, who worked as Walsingham's cipher secretary.

Phelippes was a master of his art. In a matter of days he'd cracked the cipher like a walnut and delivered the plaintext to Walsingham. The spymaster briefly considered arresting Babington and his friends, but quickly decided against it. The person he really wanted was Mary, whom he considered would be a danger to Queen Elizabeth as long as she remained alive. Walsingham figured that if Mary authorised the plot – and he had proof of it – she could be tried for treason and executed. So he filed away the plaintext in a drawer and instructed Gifford to deliver the original cipher message to Chartley Hall.

The rest of the story is one of the more painful episodes of English history. When Mary wrote back to Babington, that letter too was intercepted and deciphered. Walsingham instructed Phelippes to add an additional forged paragraph to the missive, asking Babington for the names of his fellow conspirators.

It all worked like a charm. On August 11, Mary was arrested while riding in the grounds of Chartley. Four days later, Babington and his friends were also taken into custody.

Queen Mary's trial began at Fotheringhay Castle in the Midlands on October 15 and eleven days later she was pronounced guilty of conspiring against Elizabeth, with a recommendation for the death penalty. Elizabeth hesitated for a while – it set a bad example to execute someone of royal blood: it might give people ideas – but eventually signed the warrant.

On February 8, 1587, 300 people gathered to watch the Scots Queen's beheading. It took the executioner three strokes of his axe before he severed the head completely. It was the first-ever recorded instance of decryption leading directly to the death of a monarch. The question is, how did Phelippes manage it?

9 The Arab Approach

It was the Arabs who figured how to crack ciphers that had remained mysteries for 1,000 years. What put them on the scent wasn't espionage, but religion.

In the 9th century AD, Islamic culture was the most sophisticated on the planet and there was a huge interest among scholars in discovering the exact sequence of the Prophet Muhammad's teachings. They had a theory that certain Arabic words came into being later than others, so you could work out which were the later writings of the Prophet by counting the number of the newer words in any given text.

I'm not sure if the new-word theory was actually correct, but it did lead to a lot of text analysis, studying the structure of sentences, the origins and frequency of words and, eventually, counting the number of times a particular letter of the alphabet was likely to appear on any page. They discovered, for example, that the most common letters in Arabic are A and L.

You can spot the high frequency of both letters in the lengthy name of the great Arab philosopher Abu Yusuf Ya'qub ibn Is-haq ibn as-Sabbah ibn 'omran ibn Ismail al-Kindi, who wrote nearly 300 books on subjects ranging from medicine to mathematics. In 1987, it was discovered he wrote on cryptology as well. Specifically, he put forward a brand-new method of breaking ciphers, based on the discoveries of religious scholars.

What he suggested was that if you knew the language of the original message – Arabic, English, Latin or whatever – you should grab yourself a few pages of any old text written in that language and count the number of times each letter of the alphabet appeared in it.

When you'd done with counting, you should then rank the letters in order, starting with the most frequent. This ranking became the key to cracking your cipher. You took your ciphertext and did exactly the same count there so you could rank the frequency each ciphertext symbol (or replacement letter) appeared.

Al-Kindi argued that if the most frequent symbol in your Arabic cipher was, say, ς then it was likely to stand for the letter A. If your next most frequent was Ψ, then that was likely to represent L. And so on. With a little bit of patience you'd eventually replace all the ciphertext and – bingo! – there would be your secret message in all its naked glory. The process quickly became known as *frequency analysis.*

Nowadays, of course, professional cipher-busters don't have to count the letters in a plaintext to find out their relative frequency – that job has been done for them already and the results appear in the cipher books of every Secret Service in the world. But you don't have to join MI5 to find out the frequencies in English. You can simply look them up in the table overleaf:

Alphabet Letter	Ranked Frequency	Percentage %
E	1st	12.7
T	2nd	9.1
A	3rd	8.2
O	4th	7.5
I	5th	7.0
N	6th	6.7
S	7th	6.3
H	8th	6.1
R	9th	6.0
D	10th	4.3
L	11th	4.0
C	12th =	2.8
U	12th =	2.8
M	13th =	2.4
W	13th =	2.4
F	14th	2.2
G	15th =	2.0
Y	15th =	2.0
P	16th	1.9
B	17th	1.5
V	18th	1.0
K	19th	0.8
J	20th =	0.2
X	20th =	0.2
Q	21st =	0.1
Z	21st =	0.1

That table, a bit of common sense and a lot of patience are about all you need to bust some really tricky ciphers. But before you try to put it to use, you need to take note of a couple of things., The first is that there are a few letters – C and U, M and W, G and Y, J and X, and Q and

Z – that share the same rankings. So when you make your substitutions in a ciphertext, you need to try them *both* as the replacement for any symbol that appears with the same frequency. With a bit of luck, common sense will tell you which is the right one when you've made the rest of your substitutions. For example, if your substitutions resulted in the two possibilities UOW and COW, it would make sense to assume the original plaintext was COW, since the word UOW doesn't appear in the dictionary[19].

The second thing to note is that the rankings given are all *averages*. So you can't be sure every single substitution will work. What you can be sure of is that, given enough substitutions, you should end up with sufficient information to make good sense of the cipher. This leads to the interesting fact that a long piece of ciphertext is generally easier to crack than a short one: if your secret message is really brief, the average letter rankings just don't have the chance to show up.

Keep all that in mind as you turn the page. Because in the next chapter, I'm going to show you, step by step, just how you bust a cipher.

19. Something worth knowing in your next game of Scrabble.

10 Cipher-Busting

Here's a passage of ciphertext:

RO JIUA GRMWR VAIO UJFB. NEA SAFO NOMRITA, RN WS EAB NJGAEJV TWQAI EAM R BMPT. RIOJIWJ FJJCAB RO EAM WIOAIOFY RIB NEA FJNO RVRMAIANN JS VERO VRN ERKKAIWIT.

There are 29 words and 127 letters in the message. To make life a little easier, I've actually separated out the words and given you a little punctuation, which isn't likely to happen if it was James Bond who sent you the message. I generated the cipher using a two-word keyphrase, so it isn't all that simple.

Since there is no way I'm going to tell you the keywords, even under pain of a hideous death, I'd better show you how to crack it without them. Start by counting the number of times each letter appears in the ciphertext. Here's how it works out:

A	16	J	10	R	13
B	3	K	2	S	3
C	1	M	5	T	4
E	8	N	9	U	2
F	5	O	10	V	5
G	2	P	1	W	6
I	12	Q	1	Y	1

Let's turn that into frequency rankings:

Alphabet Letter	Ranked Frequency
A	1st
R	2nd
I	3rd
O	4th =
J	4th =
N	5th
E	6th
W	7th
F	8th =
V	8th =
M	8th =
T	9th
S	10th =
B	10th =
U	11th =
K	11th =
G	11th =
Y	12th =
Q	12th =
P	12th =
C	12th =

The first thing you'll notice is that our message didn't use all the letters of the alphabet. D didn't appear at all, nor did H or L or X or Z. This is sort of interesting, but not obviously useful, except that it might suggest D, H, L, X and Z represent letters that don't occur too often in English plaintext.

But the structure of the table above soon knocks out any bright notions we might have had about simply substituting letters in our ciphertext on the basis of frequency. If you go back to the table of letter frequencies in the English language generally (it's on page 56) you'll notice it doesn't feature a joint 4th, 8th, 10th, and 11th, like the table above. And even though it does have joint 12th, the differences in the earlier rankings would obviously have knocked these out of synch.

So you can't make a straight substitution. You have to creep up on the ciphertext very subtly. All the same, the most frequent letter in the ciphertext is A and while you can't be sure that stands for the most frequent letter in the language generally (E), you might reasonably assume it could be one of the *three* most frequent letters in English, E, T or indeed, A itself. But which one?

E and A are both vowels, T is a consonant. Your English teacher won't have told you this, but vowels and consonants behave quite differently in our language. Vowels, by and large, are sociable. They like the company of almost all other letters. Consonants are a lot more picky. They'll sit beside some letters, but not others.

Just cast your eye over any page of text in this book. You'll find the vowel E turning up beside (i.e. in front of, or after) just about any other letter. A's much the same, although with a slightly narrower spread. Now look at the Ts. I doubt if there's anywhere you'll see one next to B,

D, G, J, K, M, Q or V.

So let's see how sociable our ciphertext A really is. Here's the number of times you find it beside the various other letters.

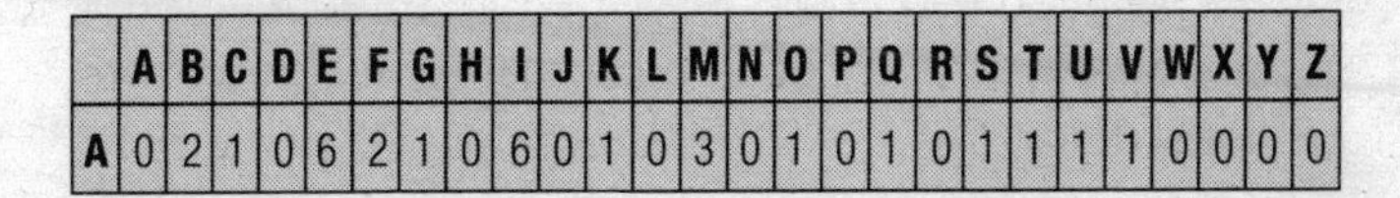

	A	B	C	D	E	F	G	H	I	J	K	L	M	N	O	P	Q	R	S	T	U	V	W	X	Y	Z
A	0	2	1	0	6	2	1	0	6	0	1	0	3	0	1	0	1	0	1	1	1	1	0	0	0	0

Let's try that for our next most frequent ciphertext letters, R, I and O.

	A	B	C	D	E	F	G	H	I	J	K	L	M	N	O	P	Q	R	S	T	U	V	W	X	Y	Z
R	0	0	0	0	2	0	1	0	3	0	1	0	3	2	3	0	0	0	0	0	0	3	1	0	0	0
I	6	1	0	0	0	0	0	0	0	2	0	0	0	0	4	0	0	3	0	1	1	0	3	0	0	0
O	1	0	0	0	0	2	0	0	4	1	0	0	1	2	0	0	0	3	0	0	0	0	0	0	0	0

At the moment, we're not trying to decide what these letters represent, just whether they're more likely to be vowels or consonants – and in particular whether cipher A is a vowel or a consonant. Remember the rule: vowels are sociable, consonants aren't. Now look at the patterns.

Cipher A has snuggled up with no fewer than 14 other characters – more than half the alphabet.

Cipher **R** sits beside 9 other letters.

Cipher **I** teams up with 8.

Cipher **O** teams up with 7.

Bear in mind you can't be absolutely sure since we're dealing with averages and the ciphertext we're examining is actually rather short, but my guess would be that A and R represent vowels while probably I and definitely O are consonants. Furthermore, since A has such a wide spread and is also the most frequent letter in our message, I'd lay good odds that it must represent the letter E.

In cipher-busting, it's always a good idea to tackle things methodically, step by step. So as a start, write down your message substituting an E for every A you find. You'll end up with this:

10

RO JIU**E** GRMWR V**E**IO UJFB. NE**E** S**E**FO

NOMRIT**E**, RN WS E**E**B NJG**E**EJV TWQ**E**I E**E**M R

BMPT. RIOJIWJ FJJC**E**B RO E**E**M WIO**E**IOFY RIB

NE**E** FJNO RVRM**E**I**E**NN JS VERO VRN

ERKK**E**IWIT.

Now we have our Es, we can start looking for the cipher letter that represents H in plaintext. The letter H comes before E quite often in the English language, but hardly ever comes immediately after it. So let's have a look at the frequency of the letters that come before and after the plaintext E we've just substituted in the message above. In the table, B stands for Before, A for After.

	A	B	C	D	E	F	G	H	I	J	K	L	M	N	O	P	Q	R	S	T	U	V	W	X	Y	Z
B	0	0	1	0	5	0	1	0	1	0	1	0	1	0	1	0	1	0	1	1	1	1	0	0	0	0
A	0	2	0	0	1	2	0	0	5	0	0	0	2	0	0	0	0	0	0	0	0	0	0	0	0	0

Look along the *Before* column in the table. One letter stands out immediately. There are five examples of the cipher letter E coming before our plaintext E in the message. In the whole wide spread of letters in this column, not another one comes close.

Now look along the *After* column. Cipher E only comes after plaintext E once (whereas cipher I comes after it five times). So we now have a letter that comes before E more frequently than any other, but hardly ever comes after it. It looks as if we've found our H. In this cipher, E = H. That's another substitution we can make:

RO JIUE GRMWR VEIO UJFB. N**HE** SEFO

NOMRIT**E**, RN WS **HE**B NJG**EH**JV TWQ**E**I **HE**M R

BMPT. RIOJIWJ FJJC**E**B RO **HE**M WIO**E**IOFY RIB

N**HE** FJNO RVRM**E**I**E**NN JS V**H**RO VRN

HRKK**E**IWIT.

This is beginning to look interesting. And now we're looking admiringly at the message, we might note that there's a single letter out on its own in Line 2:

TWQEI HEM **R** BMPT.

There are only two words of a single letter in the entire English language. One is I (capitalised) as it appears in the sentence, "I would like you to post me a million pounds, please." The other is a, as in, "A million pounds would be very nice." So the cipher R in line 2 has to be either I or A. This fits in very nicely with my earlier decision that cipher R is probably a vowel.

Store that information in the back of your head and

search the message for any two-letter words that might be there. You'll find five of them: RO (twice), RN, WS, and JS.

Every word in English contains at least one vowel. Two letter words, like those we've pulled out above, will usually contain only one vowel. (You might get a combination like Oo, as in, "Oo, look at you, cheeky thing!" but quite honestly it occurs so seldom in secret messages that you can safely ignore it.)

We already know cipher R is a vowel (I or A) so that means O and N represent consonants. It also means two out of the three letters W, J and S are vowels as well.

I know this has been pretty dull, but we're building up to a bit of a breakthrough here, so hang in. The next thing you need to look at are the three-letter words. There are seven altogether and some of them have plaintext letters already filled in.

The first one, N**HE** looks specially promising. The most common three letter words in English are *the* and *and*. So the cipher word might well be *the*. This would leave cipher N as equivalent to the letter T. But there's a problem with that. We already know that T is the third most common letter in the language and in our frequency analysis, cipher N is way down in 6th place.

So maybe N isn't T. But if not T, what else could it be? If you try every letter of the alphabet in front of HE, you'll find only two words come up – *THE* and *SHE*. We've decided T is a bit doubtful – at least I have – so let's try substituting S for N in some places and see if it starts to make sense.

RO JIU**E** GRMWR V**E**IO UJFB. **SHE** S**E**FO

SOMRIT**E**, R**S** WS **HE**B **S**JG**EH**JV TWQ**E**I **HE**M R

BMPT. RIOJIWJ FJJC**E**B RO **HE**M WIO**E**IOFY RIB

SHE FJ**S**O RVRM**E**I**E**NN JS V**H**RO VR**S**

HRKK**E**IWIT

Now let's go back to that R, which we know has to be an I or an A. Keep looking at the short words, which are the easiest and best key to any cipher. Both I and A work for RS, giving us IS or AS, but not so well for RIB. How many three-letter words can you think of that begin with I? But one of the two most common English words begins with A – *AND*.

If you substitute plaintext A for cipher R, it starts to look as if we're cooking with gas.

AO JIU**E** G**A**MW**A** V**E**IO UJFB. **SHE** S**E**FO

SOM**A**IT**E**, **AS** WS **HE**B **S**JG**EH**JV TWQ**E**I **HE**M **A**

BMPT. **A**IOJIWJ FJJC**E**B **A**O **HE**M WIO**E**IOFY

AIB **SHE** FJ**S**O **A**V**A**M**E**I**E**NN JS V**HA**O V**AS**

HAKK**E**IWIT

Grab the word **HE**M which appears twice in the cipher. Once again there are only a handful of plaintext possibilities. It could be exactly as it stands (HE**M**) or HEN, HER, HEW, HEX or HEY. All six are valid English words. But we've already deciphered the word SHE in the message (twice) which means we're talking about a female and makes the most likely decipherment of HEM to be HE**R**. How do things look if we substitute R for M throughout the cipher?

AO JIU**E** G**AR**W**A** V**E**IO UJFB. **SHE** S**E**FO **S**O**RA**IT**E**, **AS** WS **HE**B **S**JG**EH**JV TWQ**E**I **HER A** B**R**PT. **A**IOJIWJ FJJC**E**B **AO HER** WIO**E**IOFY **A**IB **SHE** FJ**S**O **A**V**ARE**I**E**NN JS V**HA**O V**AS** **HA**KK**E**IWIT

This is getting really exciting. You can see the plaintext message starting to appear right before your eyes, like a photograph in a bath of developer. But we're still not there yet, so keep concentrating on those short two/three letter words. There's one in the third line beginning with A (**A**IB) which is particularly interesting. The most frequently used three-letter word that begins with A in the English language is *AND*. Could the partly-deciphered **A**IB represent the plaintext AND? Only one way to find out. Make the substitutions and see if they work.

AO J**N**U**E** G**AR**W**A** V**EN**O UJF**D**. **SHE** S**E**FO **S**O**RAN**T**E**, **AS** WS **HED** **S**JG**EH**JV TWQ**EN** **HER A DR**PT. **AN**OJ**N**WJ FJJC**ED** **A**O **HER** W**N**O**EN**OFY **AND SHE** FJ**S**O **A**V**ARENE**NN JS V**HA**O V**AS** **HA**KK**EN**W**N**T

Uh oh – we've now got a three letter word spelled HED, which doesn't look right. Much of the rest seems sound though, so I'd let that one sit for a minute rather than throw out the whole change. Maybe just one more substitution and if that doesn't work, we'll go back a step.

At this stage I'm prepared to tackle one of the longer words, mainly because we've got one that's very close to being completely deciphered – AV**ARENE**NN, which we think is probably **A**V**ARENESS** because of the substitution made earlier. What particularly interests me is the double NN at the end. There aren't many letters that double up to end a word in English. You might think of GG, as in egg, LL as in hill, or SS, as in hiss. Of those three, the one that seems to fit best is SS, which allows you to make a really educated guess that AV**ARENE**NN is actually **AWARENESS**.

Guesses are absolutely fine for a cipher-buster. Intuition can be a very useful tool when you're faced with a difficult job. The trick is to make your guess, then make the substitutions and find out if they work. If you're wrong, no big deal: you simply try again. So let's make the substitutions here – W for V, S for all the Ns – and see what happens.

AO **JNUE** G**AR**W**A** **WEN**O UJF**D**. **SHE** S**E**FO

SO**RANTE**, **AS** WS **HED** **S**JG**EHJW** TWQ**EN** **HER**

A DRPT. **AN**OJ**N**WJ FJJC**ED** **A**O **HER**

W**N**O**EN**OFY **AND SHE** FJ**S**O **AWARENESS** JS

WHAO **WAS HA**KK**EN**W**N**T

Still holding up very nicely. Maybe that HED problem isn't a word at all, but an abbreviation with the punctuation missing. (Punctuation is frequently dropped in cipher messages.) Let's assume for the minute it's not HED but HE'D (as in *he had or he would*) and carry on.

The next obvious place to look is the word **WHA**O which is more likely to be **WHAT** than any of the few other possible alternatives. So we've now got O as the cipher equivalent of plaintext T, the third most common letter in the English language. The substitution there makes quite a difference.

AT JNUE G**AR**W**A** **WENT** UJF**D**. **SHE** S**EF**T

STRANT**E**, **AS** WS **HED** **S**JG**EH**J**W** TWQ**EN** **HER**

A DRPT. **ANT**J**N**WJ FJJC**ED AT HER**

W**NTENT**FY **AND SHE** FJ**ST AWARENESS** JS

WHAT WAS HAKK**EN**W**N**T

Now we know that cipher T = plaintext G. (It's the only possible fit for **STRANTE**.) Which gives us:

AT JNUE G**AR**W**A** **WENT** UJF**D**. **SHE** S**EF**T

STRANGE, **AS** WS **HED** **S**JG**EH**J**W** **G**WQ**EN** **HER**

A **DR**P**G**. **ANT**J**N**WJ FJJC**ED AT HER**

W**NTENT**FY **AND SHE** FJ**ST AWARENESS** JS

WHAT WAS HAKK**EN**W**NG**

That allows us to take the whole phrase **SHE** S**E**F**T** **STRANGE** and guess that it's very likely to read **SHE FELT STRANGE** in plaintext. Which means we can now substitute F for S and L for F throughout.

AT JNUE G**AR**W**A** **WENT** UJ**LD**. **SHE FELT**

STRANGE, **AS** W**F** **HED** **S**JG**EH**J**W** **G**WQ**EN** **HER**

A DRP**G**. **ANT**J**N**WJ **L**JJC**ED AT HER**

WNTENTLY **AND SHE** LJST **AWARENESS** JF
WHAT WAS HAKK**EN**W**NG**

Every time we make a new substitution, more clues arise to the rest of the message. It's a rule of cryptology that the more you decipher, the faster the whole process goes. With the last two substitutions in place, we can now return to those remaining two-letter words with a whole heap of confidence. In the second line W**F** is clearly **IF**, while further down, J**F** has to be **OF**, giving us two more important vowels. Now our message looks like this:

AT ONU**E** G**ARIA WENT** U**OLD. SHE FELT**
STRANGE, AS IF HED SOG**EHOW** G**I**Q**EN HER A**
DRPG. **ANTONIO LOO**C**ED AT HER INTENTL**Y
AND SHE LOST AWARENESS OF WHAT WAS
HAKK**EN**W**NG**

The rest is easy enough to guess. What had started out as a wholly mysterious…

RO JIUA GRMWR VAIO UJFB. NEA SAFO
NOMRITA, RN WS EAB NJGAEJV TWQAI EAM R
BMPT. RIOJIWJ FJJCAB RO EAM WIOAIOFY RIB
NEA FJNO RVRMAIANN JS VERO VRN
ERKKAIWIT.

… has now turned into the plaintext:

AT ONCE MARIA WENT COLD. SHE FELT STRANGE, AS IF HE'D SOMEHOW GIVEN HER A DRUG. ANTONIO LOOKED AT HER INTENTLY AND SHE LOST AWARENESS OF WHAT WAS HAPPENING[20].

This sort of analysis, deduction, guesswork, patient substitution and trial and error is the heart of cipher-busting, but you can sometimes short-circuit the procedure once you've found a reasonable number of plaintext letters. What you do is lay out your cipher with the known letters filled in, something like this:

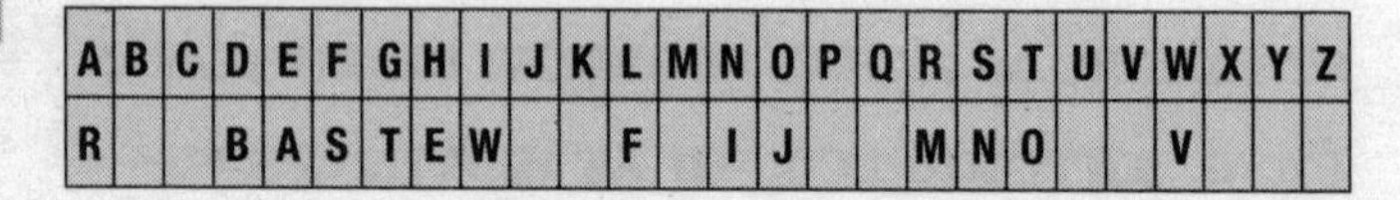

A	B	C	D	E	F	G	H	I	J	K	L	M	N	O	P	Q	R	S	T	U	V	W	X	Y	Z
R			B	A	S	T	E	W			F		I	J			M	N	O			V			

Then you take a long hard look at the second line and *see if you can guess the keyword or keyphrase.*

It's not particularly easy in the cipher I used as an example (because I'm an extremely cunning cryptologist) but even with a few letters filled in as above, you might notice that word STEW hidden in the jumble. Maybe with a few more letters, it might hit you that my keyphrase was RHUBARB STEWED (without spaces or repeated letters, of course). Once you have the keyphrase, it's easy to work out the entire cipher:

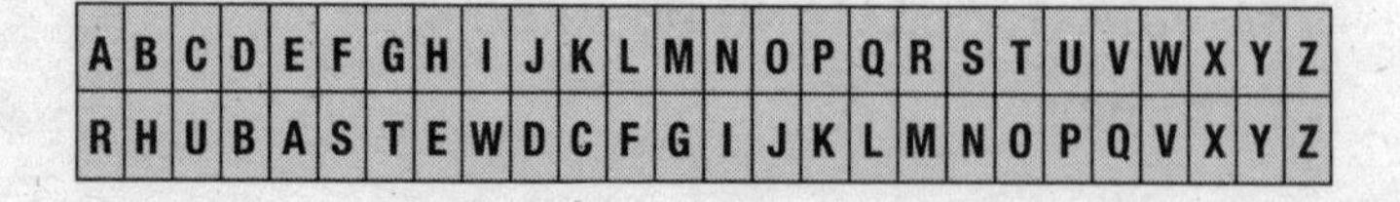

A	B	C	D	E	F	G	H	I	J	K	L	M	N	O	P	Q	R	S	T	U	V	W	X	Y	Z
R	H	U	B	A	S	T	E	W	D	C	F	G	I	J	K	L	M	N	O	P	Q	V	X	Y	Z

20. If by now you're desperate to find out what happened to Maria at the hands of the dastardly Antonio, read my book *Strange Powers of the Human Mind,* published by Faber & Faber.

And once you have the cipher, you can unlock the rest of the message without any more analysis at all. You don't *have* to find the keyphrase, as I've already shown – all it really needs is patience. But when patience is in short supply, it can be well worth shooting for the keyphrase since one good guess can save you a lot of effort.

11 Short, Sharp Ciphertest

That last chapter was heavy going, and if you worked through it along with me you really deserve a reward. So what I've done is prepare a second cipher text for you to bust all on your own.

Take your time, use everything you've learned in the last chapter, be patient and proceed step by step. The ciphertext is a lot longer than the one we worked on in the last chapter, which means frequency analysis should be more accurate. The more words you work on, the closer the letter frequencies come to the average, so there's a bit less guesswork. All the same, you're still going to have to use logic and intuition.

You don't *have* to crack this cipher, of course – there's nothing to stop you skipping this chapter altogether – but if you do tackle it, you're going to experience a very special thrill as the ciphered words gradually come clear. And talking of thrills, I should tell you at this stage that the plaintext is a passage from the *Kama Sutra*, possibly the most famous sex manual in the world. Enjoy.

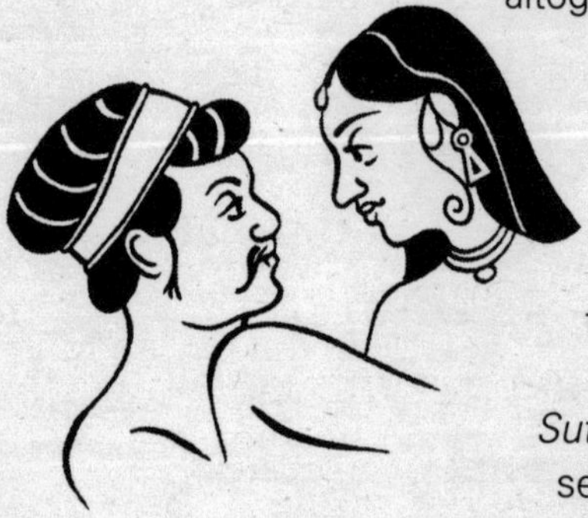

KLX UNS NLVTSNLHESR, NJWDKO OLU VP DK UNS ILRKDKO JKE PSRBLRISE NDT KSMSTTJRZ EVUDST, TNLVHE XJTN NDT USSUN, JPPHZ J HDIDUSE QVJKUDUZ LB LDKUISKUT JKE PSRBVIST UL NDT ALEZ, PVU TLIS LRKJISKUT LK NDT PSRTLK, MLHLVR NDT HDPT JKE HLLG JU NDITSHB DK UNS OHJTT. NJWDKO UNSK SJUSK ASUSH HSJWST, XDUN LUNSR UNDKOT UNJU ODWS BRJORJKMS UL UNS ILVUN, NS TNLVHE PSRBLRI NDT VTVJH AVTDKSTT. NS TNLVHE AJUNS EJDHZ, JKLDKU NDT ALEZ XDUN LDH SWSRZ LUNSR EJZ, JPPHZ J HJUNSRDKO TVATUJKMS UL NDT ALEZ SWSRZ UNRSS EJZT, OSU NDT NSJE (DKMHVEDKO BJMS) TNJWSE SWSRZ BLVR EJZT JKE UNS LUNSR PJRUT LB NDT ALEZ SWSRZ BDWS LR USK EJZT. JHH UNSTS UNDKOT TNLVHE AS ELKS XDUNLVU BJDH, JKE UNS TXSJU LB UNS JRIPDUT TNLVHE JHTL AS RSILWSE. ISJHT TNLVHE AS UJGSK DK UNS BLRSKLLK, DK UNS JBUSRKLLK, JKE JOJDK JU KDONU. JBUSR ARSJGBJTU, PJRRLUT JKE LUNSR ADRET TNLVHE AS UJVONU UL TPSJG, JKE UNS BDONUDKO LB

11

MLMGT, QVJDHT, JKE RJIT TNLVHE BLHHLX. J HDIDUSE UDIS TNLVHE AS ESWLUSE UL EDWSRTDLKT JKE UNSK TNLVHE AS UJGSK UNS IDEEJZ THSSP. JBUSR UNDT UNS NLVTSNLHESR, NJWDKO PVU LK NDT MHLUNST JKE LRKJISKUT, TNLVHE, EVRDKO UNS JBUSRKLLK, MLKWSRTS XDUN NDT BRDSKET. DK UNS SWSKDKO UNSRS TNLVHE AS TDKODKO.

(XSHH, D EDEK'U TJZ DU XLVHE AS J *KJVONUZ* PJTTJOS, EDE D?)

12 Buster Beating

Once the Arabs got cracking, no cipher was safe. You might risk a quick love letter or something similarly naughty – by the time anybody worked through the methods outlined in the last two chapters, you'd probably have broken up anyway – but for life and death messages, military matters, government secrets, it was obvious a stronger form of encryption was needed. And somebody quickly came up with it.

The new approach was ingenious, a cryptographer's form of ju-jitsu, the martial art that uses your opponent's strength against him. But before I tell you exactly what was the solution, there are a couple of things you need to know.

Up to now, every cipher alphabet in this book has involved substituting different letters of the alphabet for the letters of the alphabet in the plaintext. But this doesn't have to be the case. You can substitute *any* symbol for a letter of the alphabet and still come up with a workable cipher.

Perhaps the most convenient symbols to use would be numbers. Everybody knows them and you can often disguise them as something other than a secret message. (By pretending they're a complicated mathematical equation, for example.)

Your simplest substitution looks like this:

A	1
B	2
C	3
D	4
E	5
F	6
G	7
H	8
I	9
J	10
K	11
L	12
M	13
N	14
O	15
P	16
Q	17
R	18
S	19
T	20
U	21
V	22
W	23
X	24
Y	25
Z	26

A secret message, ciphered as follows...

20 8 9 19 9 19 1 4 21 13 2 12 1

26 25 23 1 25 20 15 3 9 16 8 5 18

... looks impressive, but wouldn't last a moment when tackled by an experienced cipher-buster. The fact that you never find a number higher than 26 tips your hand at once. It's clear numbers have been substituted for letters, almost certainly in strict sequence. Half a minute's work will confirm this and the cipher is well and truly broken.

In fact, as laid out above, it isn't a cipher at all. All you've really done is decide that certain numbers are going to represent a different alphabet. The result is absolutely equivalent to Caesar's using the Greek alphabet for his Latin message to Cicero.

That's worth remembering, since it saves you from the temptation of using shifted numbers or keyword number shifts in an attempt to pry a stronger cipher out of the table above. You *will* get a stronger cipher, of course, but it'll still be completely vulnerable to the sort of frequency analysis I described in a previous chapter. The fact that you're using numbers instead of letters *makes no difference at all.*

But what did make a difference, somewhere towards the end of the 16th century, was the realisation it was possible to screw up frequency analysis with a little ingenuity. The reasoning went like this. If E is the letter that appears most frequently in the English language – and hence shows up most often in its cipher form when encrypted – then surely the answer is to encrypt E with a whole load of *different* letters so the frequency no longer shows. You could, for example, decide that B, D, G, F, C, H, K and L *all* stood for E in your cipher. That would cut down the frequency any one of those letters appeared by *eight times*. And if you did the same with the other letters, *nothing* would be left to show up under frequency analysis.

It was a good start to beating the cipher-busters. But when you try to put it into practice, a couple of problems show themselves. The first is that to substitute more than one thing for any given letter, you can't use other alphabet letters as you did before – there aren't enough to go round. What you need is some other set of symbols, in effect a larger substitute alphabet. Numbers are ideal for this since you can use as many of them as you want.

The second problem is that to neutralise frequency analysis fully, you can't just substitute a random handful of numbers for each plaintext letter. You have to substitute *exactly the right amount* of numbers to leave the analysis null and void.

This is where the ju-jitsu came in. The cipher-makers decided to turn the cipher-buster's own research against them. In doing so, they came up with a whole new breed of cipher. Today it's called the *homophonic substitution cipher*, a term you should encrypt using any system you fancy, then forget immediately. The HSC is based on the frequency with which each letter of the alphabet appears in a typical piece of English text.

For example, if you count up 100 words of text anywhere in this book, the letter L is likely to appear about four times. (If it doesn't, try another 100-word block of text. What we're talking about here is an *average*. The more text you count, the closer to the average you'll get: trust me.) In other words, L accounts for roughly 4% of the letters in any message.

The cipher-busters had already worked out the percentages for every letter of the alphabet, since they needed that knowledge for their frequency analysis. The table they worked from looks like this:

	A	B	C	D	E	F	G	H	I	J	K	L	M	N	O	P	Q	R	S	T	U	V	W	X	Y	Z
%	8	2	3	4	12	2	2	6	6	1	1	4	2	6	7	2	1	6	6	9	3	1	2	1	2	1

(Please resist writing to tell me the figures don't add up to 100%. These are only approximate frequencies. The letter Z actually makes up *less* than 1% of appearances in English, for example, but if you rated it as zero, it wouldn't appear in your cipher at all.)

The idea that occurred to the homophonic cipher makers was that if A accounted for 8% of the letters in English text, then using eight different numbers to denote A would very effectively disguise the letter in your ciphertext, *because none of those numbers would show up under frequency analysis.*

More to the point, if you gave *every* letter the quantity of different numbers that equated to its frequency percentage, then you should end up with a completely unbreakable cipher.

Here's a table with numbers allocated, more or less at random, to each letter in the correct proportion.

A	B	C	D	E	F	G	H	I	J	K	L	M
06	15	09	02	26	13	43	90	84	10	20	86	88
18	37	17	03	64	01	27	59	34			28	12
19		16	40	29			24	25			35	
30			68	54			81	07			05	
39				46			72	73				
47				04			08	14				
57				36								
69				23								
				49								
				50								
				67								
				56								

N	O	P	Q	R	S	T	U	V	W	X	Y	Z
78	83	52	21	32	99	76	41	31	91	00	65	71
87	82	74		48	94	80	95		75		102	
53	51			93	55	92	33					
38	60			98	77	66						
44	70			85	96	45						
11	22			62	63	79						
	89					97						
						42						
						58						

You can use it to cipher a few more secret messages of your own, but in the meantime, I've used it to cipher a sample message of mine:

07 290144420 802450 2195841720

3762227587 132200 10411274 2602

70315662 800804 28397165 686027

Although that ciphertext covers just about every letter of the English alphabet – it reads, in plain, *I think the quick brown fox jumped over the lazy dog* – you won't be able to crack it by using frequency analysis. Even the ever-present E (which appeared four times in plaintext) doesn't produce a single repeated number. Which surely means you've now, at long last, got your hot little hands on the world's first truly unbreakable cipher. Well, not exactly ...

Although the HSC is an excellent, if cumbersome, way of sending secret messages that only need to get past idiots who haven't read this book, it isn't exactly unbreakable. Frequency analysis is important, but it's not everything.

Faced with my encrypted message above, a good analyst would first note that every word has an even number of numbers in it. From this, he's likely to conclude that each plaintext letter is represented by a two-figure number.

(You might confuse her, temporarily, by substituting 3-, 4-, 5- figure or longer numbers – remember the actual number is only a symbol: you can make it anything you like – but you'll quickly find you're confusing yourself as well. To be workable, your cipher has to be readable *with the key* as well as unreadable without it. As your numbers grow, your recipient has to keep looking for increasingly complex patterns of figures and you have to make sure you don't accidentally create patterns that mean something other than what you wanted to say. It all starts to be a bit of a nightmare.)

The next thing she'd do would be to home in on that single (07) number beginning the message. As we've already noted, that has to be either I or A. So she knows that 07 is one of the numbers in the I or A lists. She doesn't know which list yet, but give her time. In a longer message than my little sample, single letter words are likely to recur. When they do, she'll keep adding the results to her own lists.

And here, suddenly, frequency analysis starts to creep back. Because she knows as well as you do that, as a word, *A* is more likely to occur than *I*. So by careful comparison and a bit of logical sorting, she'll end up with every number you've allocated to A and every number you've allocated to I. Doesn't sound like much, but she'll then start to substitute Is and As throughout the message.

The next thing she'd start looking for would likely be the letter Q. In English, Q is *only* ever followed by the letter U. Since our cipher-buster knows just as much about letter frequencies as you do, she'll be well aware that Q – which turns up very seldom in your average English sentence – has only one number allocated to it, while U, which accounts for 3% of letter usage, is likely to have three. So when she finds a number in the ciphertext that doesn't appear often, but whenever it does it's always followed by one of only three other numbers, then she knows she's found her Q and, far more importantly, her U.

I'm not going to take you all the way through this one as I did before. Take my word for it, the process is tedious and time-consuming in the extreme. As a cipher, the old HSC gives a lot of trouble, but the point I want to make is that it's far from unbreakable.

And that's just one of its problems. Another is the question of decryption. To make sure your recipient will understand your message, you have to get the key to him. And in this case it's the whole key, not just a keyword or keyphrase like the ones you used in the earlier ciphers. A keyword is easy to slip through a security check, a whole cipher key sticks out like a sore thumb.

So while the homophonic substitution cipher has its moments, for the really important secret messages in your life, you really need a cipher that beats frequency analysis as effectively as the HSC, isn't as easy to crack using the word-structure method I sketched out here and allows you to communicate keys to the recipient without causing suspicion.

13 The Absolutely Positively (Nearly) Unbreakable Cipher

It was the Renaissance poet, painter, philosopher and composer Leon Alberti who first came up with one (the Renaissance was a period in Europe of great art and learning, around 1400 to 1500). Alberti figured that if one cipher alphabet still foxed most of the people most of the time, then *two* cipher alphabets (which had to be twice as good) might fox all of the people all of the time.

It was one of those ideas you just know will work. What he had in mind was a cipher constructed like this:

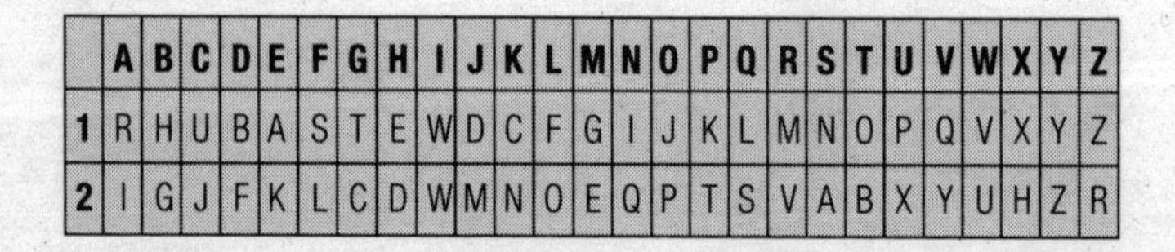

	A	B	C	D	E	F	G	H	I	J	K	L	M	N	O	P	Q	R	S	T	U	V	W	X	Y	Z
1	R	H	U	B	A	S	T	E	W	D	C	F	G	I	J	K	L	M	N	O	P	Q	V	X	Y	Z
2	I	G	J	F	K	L	C	D	W	M	N	O	E	Q	P	T	S	V	A	B	X	Y	U	H	Z	R

What you have on top is your ordinary plain bog-standard alphabet. What you have underneath are two quite different cipher alphabets, generated any way you see fit. In the table above, the first is actually my famous *rhubarb stewed* cipher from a couple of chapters ago. The second is just a more or less random arrangement of the alphabet.

As ciphers, neither of them is any better than the ones the Arabs had been cracking for centuries. But Alberti's bright idea – brilliant idea really – had nothing to do with the ciphers themselves. The whole trick was in the way you used them.

Let's suppose you wanted to encipher something short and sweet, like the word HAPPY. Using the old-style *rhubarb stewed* cipher, H would become E, A would turn into R and so on until you ended up with ERKKY. Right away you're vulnerable to frequency analysis – that repeated K is a bit of a giveaway. But imagine, Alberti thought, you *alternated* between the *two* ciphers. Imagine that the first letter of your plaintext was coded using Cipher 1, but for the next letter you jumped to Cipher 2, then went back to Cipher 1 for the third ... and so on.

If you actually try that out what you end up with (enciphering HAPPY) is EIKTZ. Right away you've lost the double K, because the P was enciphered differently each time it appeared. Even more interesting, if the P had been in a different place, it *might* have been enciphered the same way[21]. How confusing is that for any smart-ass cipher-buster?

Well, confusing ... but maybe not confusing enough. Given the information you've already read in this book you could *still* bust the cipher. It might take longer, but I have a lot of faith in your abilities. The problem is that any given plaintext letter is still encrypted using exactly the same cipher letter 50% of the time. For the other 50% of the time, it's ciphered using a different letter, but that new letter is also consistent.

What this means is that the P in HAPPY will appear

21. If you encipher a really nice name like HERBIE for example, it turns into EKMGWK, with the plaintext E turned into K both times.

throughout your ciphertext as K half the time and T the other half. You might be sharp enough to spot the pairing of the two letters, but even if you don't, the relationships between the average frequencies of certain letters in your ciphertext remain exactly the same as they always were.

Did you understand that? Doesn't matter. The point is a really experienced professional cryptographer *would* understand it and could then use the information as an iron bar to pry open the ciphertext.

Clearly, while the basic idea was a good one, it needed to be developed. But it turned out Alberti wasn't the man to do it. He just tossed out the idea and let it lie there. Fortunately several other cryptographers worked on it over the next century and one of them, a French diplomat called Blaise de Vigenère, finally produced something really neat. It's called the Vigenère Square in his honour and it looks like this:

The Vigenère Square

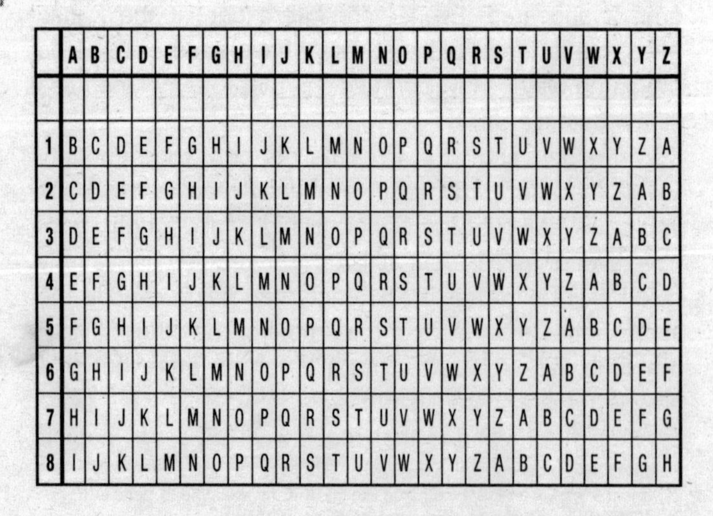

	A	B	C	D	E	F	G	H	I	J	K	L	M	N	O	P	Q	R	S	T	U	V	W	X	Y	Z
1	B	C	D	E	F	G	H	I	J	K	L	M	N	O	P	Q	R	S	T	U	V	W	X	Y	Z	A
2	C	D	E	F	G	H	I	J	K	L	M	N	O	P	Q	R	S	T	U	V	W	X	Y	Z	A	B
3	D	E	F	G	H	I	J	K	L	M	N	O	P	Q	R	S	T	U	V	W	X	Y	Z	A	B	C
4	E	F	G	H	I	J	K	L	M	N	O	P	Q	R	S	T	U	V	W	X	Y	Z	A	B	C	D
5	F	G	H	I	J	K	L	M	N	O	P	Q	R	S	T	U	V	W	X	Y	Z	A	B	C	D	E
6	G	H	I	J	K	L	M	N	O	P	Q	R	S	T	U	V	W	X	Y	Z	A	B	C	D	E	F
7	H	I	J	K	L	M	N	O	P	Q	R	S	T	U	V	W	X	Y	Z	A	B	C	D	E	F	G
8	I	J	K	L	M	N	O	P	Q	R	S	T	U	V	W	X	Y	Z	A	B	C	D	E	F	G	H

9	J	K	L	M	N	O	P	Q	R	S	T	U	V	W	X	Y	Z	A	B	C	D	E	F	G	H	I
10	K	L	M	N	O	P	Q	R	S	T	U	V	W	X	Y	Z	A	B	C	D	E	F	G	H	I	J
11	L	M	N	O	P	Q	R	S	T	U	V	W	X	Y	Z	A	B	C	D	E	F	G	H	I	J	K
12	M	N	O	P	Q	R	S	T	U	V	W	X	Y	Z	A	B	C	D	E	F	G	H	I	J	K	L
13	N	O	P	Q	R	S	T	U	V	W	X	Y	Z	A	B	C	D	E	F	G	H	I	J	K	L	M
14	O	P	Q	R	S	T	U	V	W	X	Y	Z	A	B	C	D	E	F	G	H	I	J	K	L	M	N
15	P	Q	R	S	T	U	V	W	X	Y	Z	A	B	C	D	E	F	G	H	I	J	K	L	M	N	O
16	Q	R	S	T	U	V	W	X	Y	Z	A	B	C	D	E	F	G	H	I	J	K	L	M	N	O	P
17	R	S	T	U	V	W	X	Y	Z	A	B	C	D	E	F	G	H	I	J	K	L	M	N	O	P	Q
18	S	T	U	V	W	X	Y	Z	A	B	C	D	E	F	G	H	I	J	K	L	M	N	O	P	Q	R
19	T	U	V	W	X	Y	Z	A	B	C	D	E	F	G	H	I	J	K	L	M	N	O	P	Q	R	S
20	U	V	W	X	Y	Z	A	B	C	D	E	F	G	H	I	J	K	L	M	N	O	P	Q	R	S	T
21	V	W	X	Y	Z	A	B	C	D	E	F	G	H	I	J	K	L	M	N	O	P	Q	R	S	T	U
22	W	X	Y	Z	A	B	C	D	E	F	G	H	I	J	K	L	M	N	O	P	Q	R	S	T	U	V
23	X	Y	Z	A	B	C	D	E	F	G	H	I	J	K	L	M	N	O	P	Q	R	S	T	U	V	W
24	Y	Z	A	B	C	D	E	F	G	H	I	J	K	L	M	N	O	P	Q	R	S	T	U	V	W	X
25	Z	A	B	C	D	E	F	G	H	I	J	K	L	M	N	O	P	Q	R	S	T	U	V	W	X	Y
26	A	B	C	D	E	F	G	H	I	J	K	L	M	N	O	P	Q	R	S	T	U	V	W	X	Y	Z

The top line represents the normal alphabet, of course, which you'll eventually use to help encipher your message. Below that are a series of alphabet lines, shifted like a Caesar Cipher. The first line is shifted one place (so A = B), the second two places (A = C), the third three places like the actual Caesar Cipher and so on until, on line 26, you have the plaintext alphabet all over again, unshifted.

You can obviously use the square to create shift

ciphers just like Caesar. Pick your line – avoiding 26, of course, which would only encrypt your plaintext as plaintext – create your ciphertext and send it, with the relevant line number, to somebody who has a copy of the Vigenère Square. The line number gives the key to your cipher and means your target doesn't have to work his way painfully through every shifted alphabet. But as we've already seen, this creates a cipher so weak it's hardly worth sending.

Since we know the Vigenère Square was developed on the back of Alberti's bright idea, you might be tempted to alternate the cipher lines. So a simple HELLO, ciphered using lines 1 and 2, would work out as IGMNP. And since you could pick any two lines to alternate, *including*, interestingly enough, line 26[22], the square puts a whole host of ciphers at your fingertips. Once again, you only need to let the recipient know what lines you're using and he can decipher with ease.

The only problem with this, as you've probably realised, is that because it uses simple shifted alphabets, it does not even produce as secure ciphers as the original Alberti system, let alone improve on it.

Another approach might be to cycle through three (or even more) of the lines in your Square. The first letter of your message could be ciphered using line 2, for example, the second using line 4 and the third using line 6, before you come back to using line 2 again. Now the word HELLO turns into JIRNS. You can imagine that a long message, cycling through, say, lines 2, 3, 7, 9, 11, 15, 16, 22 and 25 would begin to create real problems for a cipher-buster.

Unfortunately it's starting to produce real problems for

22. Even plaintext can be confusing. Cipher HELLO using lines 1 and 26 and you get IEMLP. Would you realise at a glance that half this one-word secret message hasn't been ciphered at all.

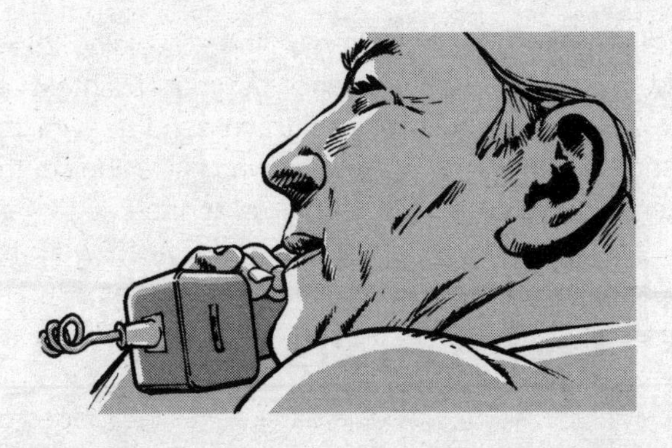

the cipher-maker as well. You need to get the sequence of lines to the recipient before he can decipher your message and, in an espionage situation, that's not easy. If your phone is tapped, you can't just call him up and say mysteriously, "Hi George, two three seven nine eleven fifteen sixteen twenty-two twenty-five nice weather we're having for the time of year." Even MI5 would be immediately suspicious. You might try pretending it's a phone number, of course, or your prediction for a winning Lotto line, but I wouldn't give much for your chances of getting away with it – and certainly not if you had to send different number sequences again and again as you would when ciphering a series of messages.

What you really need is something simpler. But that's all right because you already know about it. The secret, as before, is a keyword or keyphrase. So let me tell you a little more about the use of keywords before we tackle the Vigenère Square in its entirety.

14 Lewis Carroll's Cipher

Once you get the hang of keyword ciphers, you can get really fancy with them. One April night in 1868, the Reverend Charles Lutwidge Dodgson – Lewis Carroll to you, the bloke who wrote *Alice in Wonderland* – woke up with the idea of something he called a Telegraph Cipher. He got the name from the electric telegraph, a device patented in 1837 and widely used in those days to send messages, letter by letter, across the world[23].

Lewis Carroll

What Dodgson designed was two pieces of card, each measuring 4 x $\frac{7}{8}$ inches[24]. If you put the two of them

23. The telegraph was superseded by the telephone which meant you could talk and no longer had to send messages tediously letter by letter. Talking on the telephone has now been superseded by texting, which means you no longer have to talk and can send messages tediously letter by letter.
24. No, I don't know why he didn't make it 4 x 1 either.

together, you had a single piece of card (cut in half) measuring 4 x 1¾ inches (this was well before we converted to metric measurements), the point being you could slide one half of the card over the other. On the top half, he wrote the letters of the alphabet along the bottom edge. (He called this half his Key Alphabet.) On the bottom half he wrote the alphabet, with an additional letter A following Z, along the top edge. (He called this bottom half his Message Alphabet.) The whole device looked like this:

Key Alphabet
ABCDEFGHIJKLMNOPQRSTUVWXYZ
ABCDEFGHIJKLMNOPQRSTUVWXYZA
Message Alphabet

To use it, you agree a keyword with the recipient of your message. Find the first letter of your keyword in the Key Alphabet and the first letter of your plaintext message in the Message Alphabet. Slide the cards so these two letters line up. When you've done this, the letter in the Key Alphabet that's above the letter A in your Message Alphabet becomes the first letter of your ciphertext. Keep doing this until you have the whole ciphertext.

The pilot version of Dodgson's cipher didn't get very far[25], but he worked hard to improve it and finally came up with something really neat. This is what it looked like:

25. You can see why if you've tried to follow my instructions.

Lewis Carroll's Cipher

	A	B	C	D	E	F	G	H	I	J	K	L	M	N	O	P	Q	R	S	T	U	V	W	X	Y	Z	
A	a	b	c	d	e	f	g	h	i	j	k	l	m	n	o	p	q	r	s	t	u	v	w	x	y	z	A
B	b	c	d	e	f	g	h	i	j	k	l	m	n	o	p	q	r	s	t	u	v	w	x	y	z	a	B
C	c	d	e	f	g	h	i	j	k	l	m	n	o	p	q	r	s	t	u	v	w	x	y	z	a	b	C
D	d	e	f	g	h	i	j	k	l	m	n	o	p	q	r	s	t	u	v	w	x	y	z	a	b	c	D
E	e	f	g	h	i	j	k	l	m	n	o	p	q	r	s	t	u	v	w	x	y	z	a	b	c	d	E
F	f	g	h	i	j	k	l	m	n	o	p	q	r	s	t	u	v	w	x	y	z	a	b	c	d	e	F
G	g	h	i	j	k	l	m	n	o	p	q	r	s	t	u	v	w	x	y	z	a	b	c	d	e	f	G
H	h	i	j	k	l	m	n	o	p	q	r	s	t	u	v	w	x	y	z	a	b	c	d	e	f	g	H
I	i	j	k	l	m	n	o	p	q	r	s	t	u	v	w	x	y	z	a	b	c	d	e	f	g	h	I
J	j	k	l	m	n	o	p	q	r	s	t	u	v	w	x	y	z	a	b	c	d	e	f	g	h	i	J
K	k	l	m	n	o	p	q	r	s	t	u	v	w	x	y	z	a	b	c	d	e	f	g	h	i	j	K
L	l	m	n	o	p	q	r	s	t	u	v	w	x	y	z	a	b	c	d	e	f	g	h	i	j	k	L
M	m	n	o	p	q	r	s	t	u	v	w	x	y	z	a	b	c	d	e	f	g	h	i	j	k	l	M
N	n	o	p	q	r	s	t	u	v	w	x	y	z	a	b	c	d	e	f	g	h	i	j	k	l	m	N
O	o	p	q	r	s	t	u	v	w	x	y	z	a	b	c	d	e	f	g	h	i	j	k	l	m	n	O
P	p	q	r	s	t	u	v	w	x	y	z	a	b	c	d	e	f	g	h	i	j	k	l	m	n	o	P
Q	q	r	s	t	u	v	w	x	y	z	a	b	c	d	e	f	g	h	i	j	k	l	m	n	o	p	Q

R	r	s	t	u	v	w	x	y	z	a	b	c	d	e	f	g	h	i	j	k	l	m	n	o	p	q	R
S	s	t	u	v	w	x	y	z	a	b	c	d	e	f	g	h	i	j	k	l	m	n	o	p	q	r	S
T	t	u	v	w	x	y	z	a	b	c	d	e	f	g	h	i	j	k	l	m	n	o	p	q	r	s	T
U	u	v	w	x	y	z	a	b	c	d	e	f	g	h	i	j	k	l	m	n	o	p	q	r	s	t	U
V	v	w	x	y	z	a	b	c	d	e	f	g	h	i	j	k	l	m	n	o	p	q	r	s	t	u	V
W	w	x	y	z	a	b	c	d	e	f	g	h	i	j	k	l	m	n	o	p	q	r	s	t	u	v	W
X	x	y	z	a	b	c	d	e	f	g	h	i	j	k	l	m	n	o	p	q	r	s	t	u	v	w	X
Y	y	z	a	b	c	d	e	f	g	h	i	j	k	l	m	n	o	p	q	r	s	t	u	v	w	x	Y
Z	z	a	b	c	d	e	f	g	h	i	j	k	l	m	n	o	p	q	r	s	t	u	v	w	x	y	Z

Are you seized by *déjà vu*? Do you have the weird feeling you've been here before? You're absolutely right. Lewis Carroll's cipher isn't identical to the Vigenère Square, but it comes within a whisker. I'm not saying the reverend gentleman stole his bright idea from Vigenère, but there's no doubt at all that if you learn how to use this cipher, you're equipped to use the standard Vigenère Square.

When you look at the grid above, there's not a hint of a keyword in sight, but it's a keyword cipher just the same. Both you and the recipient of your secret messages must have a copy of the cipher itself[26]. You also have to agree a keyword, or keyphrase, between you. (Obviously you can change the keyword/keyphrase as often as you like if you're sending a whole series of messages.) Now here's how you use it.

26. Which is a really good reason for buying *two* copies of this book.

Start by composing your message. We'll suppose it's along the lines of:

Let's kidnap Fred and eat his brains.

Write it down in plaintext with any spaces, punctuation and capital letters removed, like this:

letskidnapfredandeathisbrains

Now take your agreed keyword – for example *braineaters* – and write it above the plaintext. Repeat it as often as you need so every letter of the plaintext is covered. The result should look something like this:

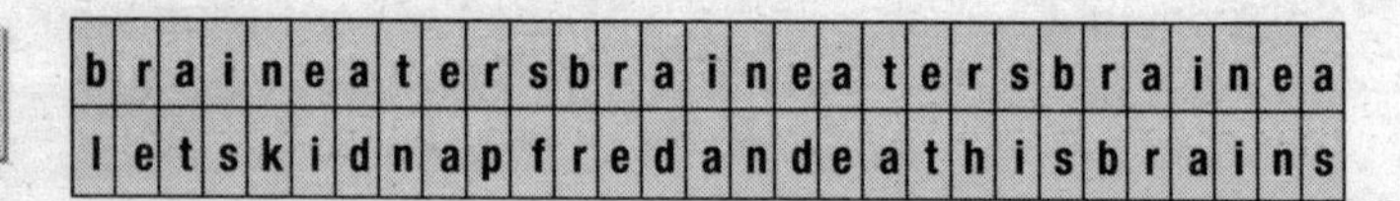

b	r	a	i	n	e	a	t	e	r	s	b	r	a	i	n	e	a	t	e	r	s	b	r	a	i	n	e	a
l	e	t	s	k	i	d	n	a	p	f	r	e	d	a	n	d	e	a	t	h	i	s	b	r	a	i	n	s

You're now ready to encipher the message. Take Lewis Carroll's table. You'll notice there's a complete alphabet along the top and two others running down the left and right hand sides. Run your finger along the top until you come to the first letter of your keyword. (You won't have to run very far using my example since the first letter of the keyword is B.)

The letter directly beneath the first B of your keyword is L. Run your finger down the alphabet on the left hand side of the table until you reach L. Now move across until it's in a line with the B you identified along the top. Your finger should now be resting on the lower-case letter 'm.' That becomes the very first letter of your cipher message, replacing the 'l' of 'lets' in the plaintext.

Instructions like this are hell to write, but you'll see what I'm getting at if you actually try it. Think of Carroll's table as having *columns* (running up and down) and *rows* (running left to right.) Every column is labelled by a letter of the alphabet running along the top. Every row is labelled by a letter of the alphabet running down the side.

To cipher your message, all you need to know is that each letter of your keyword refers to a column, while each letter of your plaintext message refers to a row. Bring the two together and you have your ciphertext.

Let's take it a bit further with our example. You already know the first letter of your ciphertext is *m*. Write that down on a new line underneath the *l* of your plaintext. The next letter of your keyword is *r* with the letter *e* in plaintext underneath it. Go to Column *R* along the top and run your finger down until you reach Row E. The table will show you that the *e* of *lets* should become a *v*. Write that down as the second letter of your ciphertext. After you work carefully through the rest of your message, you'll end up with this:

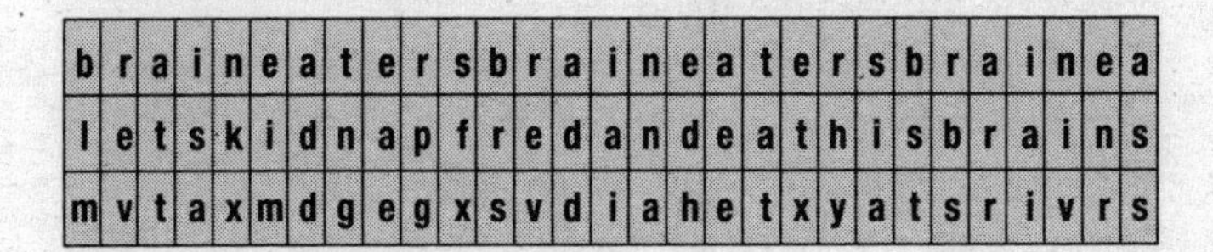

b	r	a	i	n	e	a	t	e	r	s	b	r	a	i	n	e	a	t	e	r	s	b	r	a	i	n	e	a
l	e	t	s	k	i	d	n	a	p	f	r	e	d	a	n	d	e	a	t	h	i	s	b	r	a	i	n	s
m	v	t	a	x	m	d	g	e	g	x	s	v	d	i	a	h	e	t	x	y	a	t	s	r	i	v	r	s

Now throw away the first two lines and send your ciphertext message:

Mvta xmdgeg xsvd iah etx yat srivrs![27]

You use the classic Vigenère Square in exactly the same way. First write your keyword (repeatedly if necessary) above your plaintext message, just as you did with *braineaters*. So if your secret message is HELLO, HELLO, it will look like:

BRAINEATERS
HELLO HELLO

To cipher the H, you look down the left-hand column of the Square until you find the letter B. As it happens, you won't have far to look – it's in row 1. You look along that row until you come to the letter under plaintext H and discover it's I, which becomes the first letter of your ciphertext. Then you start looking for R in the lefthand column. When you find it at the start of row 17, you read across to note the letter under E is V, which is the next letter of your ciphertext. And so it goes until you've completed the whole thing.

The really interesting thing is that while you've ciphered the same word twice, the results look nothing like one another. In the example above, the first HELLO comes out as IVLTB while the second now reads HXPCG. You've not only suppressed the letter pattern of the double L and the repeat of any given word, you've

27. To translate it back, you look in the B column for the letter M, then look sideways to discover it's in row L. This gives you the first letter of the plaintext. Then keep working through your keyword, letter by letter, until you've deciphered the whole message.

managed to suppress the logical repetition of any letter whatsoever, something that works whatever keyword or keyphrase you select.

Clever or what? Some of the plaintext letters are ciphered *as themselves*. Some of them have a different cipher letter *depending on their position in the message*.

You'd imagine that without the keyword nobody, but nobody, could ever crack a cipher as cunning as that. But you'd be wrong.

15 Square Bashing

It was a Victorian gent named Charles Babbage who finally bashed the Vigenère Square[28]. Babbage came up with a method of decryption so clever – and so complicated – I'm not sure I can explain it to you clearly. But I'm going to give it a try.

At first glance, the Vigenère Square leaves you with the impression your plaintext could be encrypted using just about any combination of cipher alphabets under the sun. And in theory that's more or less true. But Babbage suddenly realised it wasn't true in practice. Because as soon as you decided on your keyword, you automatically *limited* the options available for the Vignère Square.

Take a simple piece of plaintext for example that starts. *The cat sat on the mat* and goes on for 40 pages. In theory there are thousands of different ways you could encrypt each word. But once you pick your keyword – let's say DUCK for this example – it can only be encrypted *four ways.* Here's how that works: First, you tag your keyword to your plaintext.

D	U	C	K	D	U	C	K	D	U	C	K	D	U	C	K	D
T	H	E	C	A	T	S	A	T	O	N	T	H	E	M	A	T

28. You may have heard of him already. He was the man who invented the world's first computer. He called it a 'Difference Engine' and it contained 25,000 precision parts, mainly made from brass.

Now look at the beginning letter of your message, which happens to be T. The first time it's ciphered using the Vigenère Square, it turns into W, because it's encrypted using the D of DUCK. The second time you hit it, it turns into N, because it happens to fall under the U of DUCK. If it links with the C of DUCK later in the message, it'll be ciphered as V, while the K of DUCK will turn it into D. But with DUCK as your keyword, those are your only possible options. (I know you can have longer and more complex keywords, but I'm keeping this simple to show you the *principle* Babbage stumbled on.)

What goes for single letters, goes for whole words as well: they can only be ciphered so many different ways, depending on where they fall under your keyword. Let's suppose the bloke who wrote your plaintext was afflicted with a stutter so that it came out *The the the the the*. Ciphering it using our DUCK key would result in:

D	U	C	K	D	U	C	K	D	U	C	K	D	U	C	K	D
T	H	E	T	H	E	T	H	E	T	H	E	T	H	E	T	H

WBG DKY VRH NJO WBG DKY etc

Look carefully at that ciphertext. The first four words are fine, but after that it *starts to recycle*, ciphering the same word in the same (four) ways over and over. That becomes pretty obvious if your plaintext consists of the same repeated word. Where Babbage showed his brains was realising the Vigenère Square has to recycle *everything* eventually *and that the frequency with which it recycles tells you the length of the keyword.*

You can see that last bit clearly in our stuttering example. The same sequence of cipher letters comes up every four words, showing that the keyword has to be four letters long. (If you'd used the codeword DUCKLING, the same ciphers would turn up every *eight* words, because DUCKLING is eight letters long.) From this interesting observation, Babbage worked out his whole system of Vigenère Square bashing. Here's how you do it:

- First you look for letter sequences that appear more than once in the ciphertext (like WBG in the example above).

- Assume these sequences have been ciphered using the same part of the keyword. (This doesn't actually have to be the case. You *could* get a repeated sequence purely by accident, but the chance of that happening is small: and the longer the sequence of letters the smaller the chance gets. Frankly, if you're dealing with a five-letter sequence, you can safely assume it's keyword coded and I'd personally take my chances with a four-letter or even three-letter sequence.)

- Count the number of letters between each repetition. You have to be careful how you do this. In the (nonsense) ciphertext PQRVSTWXPMNIPQRV, the sequence PQRV recycles after 12 letters. You start counting with the second letter of the sequence (in this case Q, which = 1) and end with the next appearance of the first letter of the sequence (in this case P which is 12). Keep a careful note of your results.

- I hate to do this to you, but you now have to work out the *factors* of the various numbers you've just noted down. If you'd stayed awake in maths class, you'd know that a factor is any number that divides into another number evenly (without leaving a remainder). If you take the number 20, for example, its factors are 1, 2, 4, 5, 10 and 20 itself, because all these numbers divide evenly into 20 without leaving a remainder.

- For the purpose of this exercise, you can forget the number 1 as a factor completely, because that would point to a 1-letter keyword which will always produce a really dumb ciphertext. Try it. You'll discover you're right back there with old Julius Caesar, creating a shift cipher any idiot could crack.

- Compare the different factors for the various numbers you've collected until you find one that's common to them all. *This will almost certainly give you the number of letters in your keyword.*

- Suppose, for example, the spaces between recycling the first letter sequence was 85, the second was 20, the third was 110 and the fourth was 5. The factors of 85 are 1, 5, 17 and 85. The factors of 20 are 1, 2, 4, 5, 10 and 20. The factors of 110 are 1, 2,5, 10, 11, 22, 55, and 110. The only factors of 5 are 1 and 5[29]. The only factor common to all those (except for 1, which you ignore) is 5. Which means your keyword is five letters long.

29. That's assuming I've got it right. (I already warned you I'm no good at maths.) But even if I'm wrong, it's only an example.

- Now you can start cooking with gas.
- Now you can start hunting for the keyword.
- Just to remind you, the Vigenère Square consists of several rows of different cipher alphabets. Any given one of them is a straightforward shifted Caesar type cipher, dead simple in itself. You know that one of the rows of the Vigenère Square – that's to say, one of these simple shifted alphabets – has been used to encrypt the first, sixth, eleventh, sixteenth, twenty-first (and so on) letters of your ciphertext. *And you can use frequency analysis to figure out which row.*
- You do that by graphing the frequencies of the 1st, 6th, 11th, 16th, 21st, etc letters of your ciphertext, then comparing it with a similar frequency graph created using a passage of English plaintext of the same length as your ciphertext. Note the places where the two graphs look alike (they won't be identical unless you're very lucky) and this will point directly to the number of places the alphabet was shifted to give you the row you need in the Vigenère Square.
- The other way of doing it is to search for Ts and Es using frequency analysis and use the results to find the displacement. This isn't as accurate as looking at graphed patterns so you'll have to check your results by trial and error, but with luck and patience you'll discover the right row in the end.
- Once you know the row, you only have to look across to find the first letter of your keyword. (Exciting, eh?)

- Now repeat the same trick with the remaining four letters and you're left with the entire keyword. With the entire keyword, the ciphertext is cracked. (And actually, a lot of the time you don't have to work out the keyword or keyphrase laboriously letter by letter. Keywords usually make sense so they're easy to pass on, which means having got some of the letters you can often guess at the rest. If you discovered the first three letters were DUC, for example, you could be tempted to guess the fourth might be K.)

This is a frighteningly clever, subtle and sophisticated system for breaking into a Vigenère Square and if you found it difficult to follow, imagine what a nightmare it was to describe. But for all its cleverness, subtlety and sophistication, there are both stronger and weaker ciphers you can't break this way. We'll be examining a few of them in the remainder of this book.

16 The Wheatstone Digraph

While Babbage was busy cracking the Vigenère Square, one of his closest friends, Sir Charles Wheatstone, was equally busy inventing a new type of cipher he hoped would prove even tougher.

This one, like so many others, starts with an agreed keyword. Since one of the things you'll probably use ciphers for is sending naughty messages, let's assume your keyword in this instance is NAUGHTY.

The first thing you do to create a Wheatstone Cipher[30] is write down the letters of the alphabet in a 5x5 square, starting with the keyword. I selected the example keyword rather cleverly so there were no repeated letters, but if your keyword is something like MOON or BIBLE, then you have to strike out any repetitions, so it becomes MON or BILE.

By now you've probably realised that a 26-letter alphabet won't fit into a 5 x 5 square, which has only 25 spaces. To get around this, you decide to view the letters I and J as one and the same. (If this confuses code-busters, so much the better.)

This done, you can generate a Wheatstone Square.

30. Actually it's now generally known as the Playfair Cipher since it was popularised by Baron Playfair, but I like to give credit to the man who invented it.

Using the keyword NAUGHTY, my example square looks like this:

N	A	U	G	H
T	Y	B	C	D
E	F	I/J	K	L
M	O	P	Q	R
S	V	W	X	Z

Now write down your plaintext message. For the present example, mine will be:

FRESH HAGGIS IS ONE OF MY
FAVOURITE FOODS[31].

Before you can encrypt it, you need to break it into digraphs, a fancy way of saying *paired letters*. The cipher won't work unless the letters of each digraph are different, so when you get doubles, like the GG in HAGGIS or the OO in FOODS, you have to split them up with an X. And if your message happens to have an odd number of letters, you need to add an X after the last one to make it into a digraph. All of which leaves my plaintext looking like this:

FR ES HX HA GX GI SI SO NE OF MY
FA VO UR IT EF OX OD SX

It's beginning to look encrypted already, but you ain't seen nothing yet!

31. Which has the added benefit of being true.

Sir Charles came up with a set of rules that governed the final encryption once you reached this stage. You start by taking the first digraph (FR in the example) and searching for each of its two letters in your Wheatstone Square. What you're trying to find out is whether both letters fall in the same row or possibly in the same column. With FR it's neither. The F falls in row three (counting from the top) and column two (counting from the left). The R falls in row four, column five.

Charles Babbage

What you do to encrypt the digraph depends on the position of its letters. Since my first two aren't in the same row or column, I look along the *row* containing the F until I reach the *column* containing the R. The letter that marks the spot where this row and that column intersect becomes the cipher for the first letter of my digraph. If you check the Wheatstone Square, you'll discover that letter is L.

I follow the same process to encrypt the second letter of my first digraph. Looking along the row that contains R, I reach the column that contains F and find the letter at the intersection is O. My first digraph, FR in plaintext, is encrypted as LO.

Onwards and upwards. My next digraph is ES. These

two letters aren't in the same row, but they *are* in the same column. The rule for letters in the same column is that each is replaced by the letter immediately below it. Thus – look at the square – E is replaced by M. The S is a bit more tricky since it lies at the bottom of the column, so there's no letter underneath it. But in a situation like that, you simply replace it with the letter at the very *top* of the column – in this case N. So now my cipher message is looking like this: LO MN ...

The next two plaintext letters are HX. They don't fall in the same row or column, so they encrypt as GZ.

There's a change of rule for the next two letters HA, which happen to fall in the same *row*. When letters fall in the same row, they're each replaced by the letter on the immediate right. Unfortunately, the letter H is at the very end of its row so there isn't a letter on its right, but in that case you replace with the letter at the beginning of the row. So H becomes N. The next letter of the digraph, A, is straightforward. It turns into U, the letter on its right.

At this point, you have all the rules you need to use the Wheatstone Cipher. If the letters of your digraph are in the same column, you replace each with the letter underneath. If they're in the same row, you replace each with the letter on its right. And if they're in different rows and columns, you look along the row until you reach the junction with the column that contains the other letter and replace with the letter you find there.

By following these rules, my haggis message encrypts as:

LO MN GZ NU CG UK WE NV TM VO OT OY AV
HP EB FJ QV RY VZ

This is a nifty little cipher – one of my personal favourites since it's simple to learn, quick and easy to use. But it has a basic weakness. It's susceptible to the old cipher-buster's trick of frequency analysis; although a very slightly different form of frequency analysis to the one you learned earlier. If you ever find yourself faced with a solid block of text encrypted with the Wheatstone Cipher, do a careful count to find the most frequent digraphs. When you've done that, try replacing them with the most frequently paired letters in the English language.

TH	HE	AN	IN	ER	RE	ES

17 We Have Ways of Making a Cipher

Although the Wheatstone Cipher was taken up by the British War Office and used in the Anglo-Boer War (the end of the 1900s), its weakness soon began to show. As a result, it was abandoned and the world waited breathlessly for somebody to invent a new method of encryption that would – this time – prove to be unbreakable.

And in March, 1918, in marched the Germans with one they'd engineered earlier. They were convinced their invention could not be cracked by any method known to man. With their well-known talent for a catchy title, they called it the ADFGVX Cipher.

To make up your own version of the ADFGVX Cipher, you start by drawing up a 7 x 7 square grid, like this:

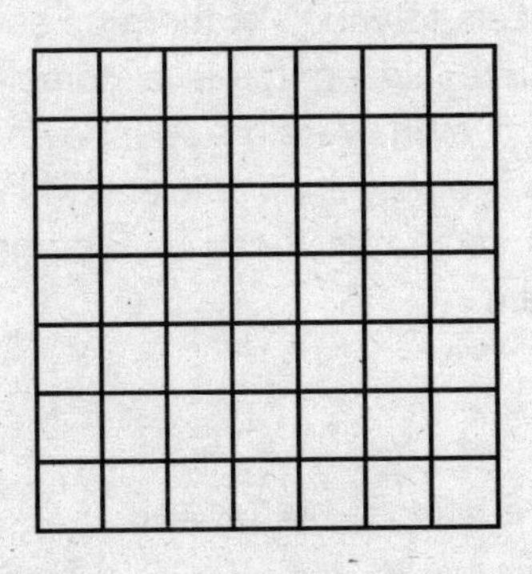

In the top row of squares, you write the letters ADFGVX, starting in the second square, then do the same down the first (left hand) column, again starting at the second square[32]. This leaves your grid looking like this:

	A	D	F	G	V	X
A						
D						
F						
G						
V						
X						

If you're one of those people who counts everything, you'll have noticed there are now 37 empty squares left, or 36 if we ignore the one that's out on its own in the top left-hand corner. Thirty-six squares is sort of interesting, since it allows you to write in all the letters of the alphabet, plus the basic digits 0 to 9.

Which is exactly what I want you to do now in any order that appeals to you. Mix them up as much as you like. What you're aiming for is a completely random sequence. My own version is shown below, but remember this is purely an example: if you want to make your cipher as secure as it can be, you need to create your own square.

17

32. Yes, that's why they called it the ADFGVX Cipher.

	A	D	F	G	V	X
A	Q	5	U	I	9	0
D	6	W	J	K	L	8
F	P	0	E	A	S	1
G	D	F	G	R	4	H
V	Z	X	C	2	T	V
X	7	B	N	M	3	Y

Now we need the secret message you want to cipher. Let's suppose it's:

NOW WE NEED THE SECRET MESSAGE YOU WANT TO CIPHER.

You begin by exchanging each letter of the message for the coordinates that mark its position on the grid. You give the column coordinate first, followed by the row, so that first N becomes XF. Searching out the rest of the letters one at a time turns the message into:

XF AX DD DD FF XF FF FF GA VV GX FF FV FF VF GG FF VV XG FF FV FV FG GF FF XX AX AF DD FG XF VV VV AX VF AG FA GX FF GG

If the message happens to contain a number – like NOW WE NEED THE 1 SECRET MESSAGE YOU WANT TO CIPHER – you'd encrypt that just the same as the letters. Thus 1 would become FX and a number like 223875 would become VG VG XV DX XA AD.

So far this isn't exactly the world's safest cipher. If you were trying to crack it, you'd already be eyeing all those

FFs with suspicion, reckoning (quite correctly) they most likely stood for E. But we haven't finished yet. Your next job is to draw another grid, this time based on an agreed keyword. I'm going to use the keyword MOUSE since that's the name of one of my cats[33]. You can use any keyword you like. The only thing to remember is that you're going to have to pass both the keyword and your own initial grid on to whoever is going to receive your secret message.

Since *mouse* has five letters, my grid will have five columns and as many rows as is needed to accommodate each of the letters in the first stage of my cipher message. (Which will be twice the number of letters there were in the plaintext, as each letter has now encrypted into *two* coordinate letters.) Using the example message above, my grid fills in like this:

M	O	U	S	E
X	F	A	X	D
D	D	D	F	F
X	F	F	F	F
F	G	A	V	V
G	X	F	F	F
V	F	F	V	F
G	G	F	F	V
V	X	G	F	F
F	V	F	V	F
G	G	F	F	F
X	X	A	X	A
F	D	D	F	G
X	F	V	V	V
V	A	X	V	F
A	G	F	A	G
X	F	F	G	G

33.One of my more confused cats. I've often wondered if there's a connection.

Now you rearrange the letters of your keyword into alphabetical order. Thus MOUSE becomes EMOSU. Then you rearrange the columns of your grid so they follow the newly arranged MOUSE, like this:

E	M	O	S	U
D	X	F	X	A
F	D	D	F	D
F	X	F	F	F
V	F	G	V	A
F	G	X	F	F
F	V	F	V	F
V	G	G	F	F
F	V	X	F	G
F	F	V	V	F
F	G	G	F	F
A	X	X	X	A
G	F	D	F	D
V	X	F	V	V
F	V	A	V	X
G	A	G	A	F
G	X	F	G	F

To create your final cipher, all you have to do is go down each column in turn and write out the letters as they appear. This makes my secret message come out as:

DF FV FF VF FF AG VF GG XD XF GV GV FG XF
XV AX FD FG XF GX VG XD FA GF XF FV FV
FF VF XF VV AG AD FA FF FG FF AD VX FF

To decrypt it, all you have to do is reverse the steps I took to cipher it – that's to say, you arrange the keyword (which you've already received) into alphabetical order, draw a grid, fill in the letters of the message, column by column in order, rearrange the columns back into the order they appeared in the keyword, then start to decipher using the letters to find your plaintext coordinates in the original grid (which you should also have received.)

By the time you've done that, you'll probably have decided that without keyword and original grid there's no way on God's green earth that cipher could be cracked. In fact, it took a French cryptographer named Georges Painvin less than a month to bust it open. But he lost a lot of weight in the process.

18 The Absolutely Positively Unbreakable Cipher (Honest)

By this stage you must be wondering if a completely unbreakable cipher actually exists. Once the Vigenère Square was cracked a lot of cryptographers were thinking the same way. But that didn't stop them trying.

For all the fancy tricks that came up after Vigenère, the good old-fashioned bog-standard keyword cipher still seemed to hold out the best hope. Was there, perhaps, some way of improving it? And not by complicating things like Vigenère did, but by tinkering with the very heart of the cipher so its foundations were strengthened.

To find something's strength, it's often a good idea to look at its weaknesses. The essential weakness of a keyword cipher, in whatever form it appeared, had always been the fact that it recycled keywords. That's what helped Babbage crack the Vigenère Square. That's what left every other keyword cipher vulnerable. Take our earlier braineaters example:

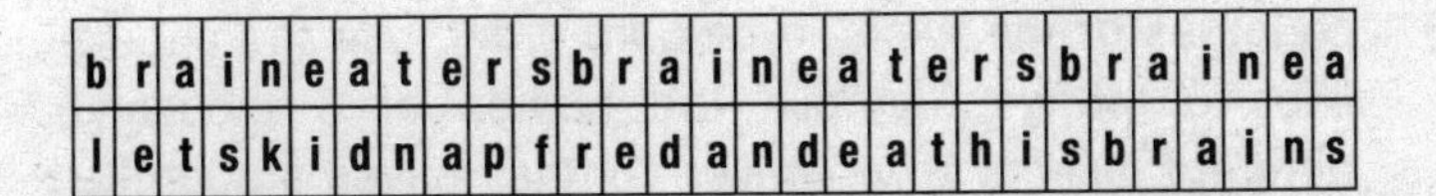

b	r	a	i	n	e	a	t	e	r	s	b	r	a	i	n	e	a	t	e	r	s	b	r	a	i	n	e	a
l	e	t	s	k	i	d	n	a	p	f	r	e	d	a	n	d	e	a	t	h	i	s	b	r	a	i	n	s

Although *braineaters* is a lengthy and unusual keyword, the cipher still recycles after every twelfth use. Even if you used the longest single keyword in the English language – *antidisestablishmentarianism* – you'd still have to recycle on the 29th usage. Most secret messages are a lot longer than the examples in this book, but that makes them easier to crack, not more difficult. If you have a 2,000-word message, enciphered with a 10-letter keyword, the Babbage method leaves you with just 10 sets of 200 words each to analyse for frequencies. Which may take a little time, but it's not going to give you any serious trouble.

All the same, what would happen if you used a key that was literally as long as the message itself? Now that's a really interesting thought. Even though the longest word in English is just 28 letters long, you're not stuck with a 29-cycle keyword. As we noted in passing earlier, there's nothing to stop you using a keyphrase. A key*phrase* can be as long as you like. And that means it can be as long as your original message. Here's how a keyphrase might look if we were still hunting for poor Fred's brains:

W	H	Y	S	H	O	U	L	D	A	B	A	L	D	M	A	N	P	A	I	N	T	R	A	B	B	I	T	S
l	e	t	s	k	i	d	n	a	p	f	r	e	d	a	n	d	e	a	t	h	i	s	b	r	a	i	n	s

For a longer message that required a longer keyphrase, you could pick a page or two from a book, list the countries of the world in random order, use the chorus lines of songs, compile the names of a thousand different insects or whatever. All that leaves a potential cipher-buster with a real problem. If your message is

2,000 words long, enciphered with a 2,000-word keyphrase, then instead of analysing 10 sets of 200 words (which is easy) he has to analyse 2,000 sets of one word each (which is impossible).

So when the keyphrase is the same length as the message, the cipher can't be broken – you'd wonder why nobody ever thought of it before. Or maybe you wouldn't. Because while it pains me to tell you this, the cipher can still be broken. It just can't be broken using the traditional method of frequency analysis.

In order to crack a cipher of this type, you start by assuming two things. The first is that the keyphrase used makes sense, or at least contains words in plain English. The second is that the word THE must be included in the plaintext somewhere[34].

What you do then is pick a few examples of three-letter sequences throughout the ciphertext and pretend each one represents the word THE. Use each one in turn to find out possible bits of the keyphrase. What you're looking for is keyphrase fragments that seem to be sensible bits of English words. For example, the sequence ING appears quite often in English, but a sequence like PXZ is just a mess.

As you start to pick up fragments of the keyphrase, you can begin to guess whole words and test them by using them to decipher more of the message. If it works, keep going. If it doesn't, substitute a few more THEs (or other common words) and go again.

This, believe me, is a brutally tedious system based on guesswork and trial and error that could give you days of mind-numbing effort in decrypting a message of even moderate length, but it *does* work. So while a keyphrase the same length as the message produces a really strong

34. Or AND or BUT or any common word, really.

cipher, it's not actually unbreakable. But a slight variation of this cipher system really is.

What left your last cipher vulnerable was one thing and one thing only – the fact that your keyphrase made sense; or at very least was composed of *meaningful* words. A very different situation arises if you use a random key; that's to say a key (the same length as your message) composed entirely of a random arrangement of letters making no sense whatsoever.

If you do this and never use the same key twice (because repeating it for message after message will start to generate patterns that a good analyst will use against you) then you have a cipher that's guaranteed unbreakable.

Think it through. If a prospective cipher-buster tries substituting THEs at random, the segments of keyword he discovers will *never* make sense so he'll never discover when he's right. Frequency analysis won't work, of course – he couldn't even use that method *before* you tried a random keyword – so his only remaining option is to test every possible keyword by trial and error.

When the idea of a random keyword was first put forward, towards the end of World War One, testing every key by trial and error was physically impossible. Even a short message generates multi-millions of possibilities and an analyst would be dead before he could try them all. Today you might manage to do the job within a worthwhile timescale using a supercomputer, but that still wouldn't give you the original plaintext.

The problem is that checking out every possible combination of key generates every possible message that can be created within a given length of text. And there's absolutely no way of knowing which of these messages is the real one.

Is there a problem with this approach, as there have proved to be problems with every other approach so far? Well ... yes and no. The cipher is unbreakable: you can take my word for that. But you have to make sure you generate an absolutely random key. That's not quite as easy as it sounds. If you were to try typing letters purely at random you'll soon find you fall into quite unconscious patterns; and once a pattern creeps in, however subtly, you compromise your cipher.

And assuming you do find a way to generate a totally random key – which is possible, just not especially easy – then you have to find a way to distribute it.

The US Army officer, Major Joseph Mauborgne, who came up with the random key idea, advocated what he called the *one-time pad* approach. You printed up thick pads containing hundreds of pages of random letters to be used as keys. Only two copies of each pad would be made, one for the sender of the secret message, one for the receiver. Each page would be used only once, then thrown away[35].

One-time pads worked, but only up to a point. Once a war started, there were so many secret messages flying around that they became impractical. The problem of printing up and securely distributing thousands of them became a nightmare. The military cast around for something that would speed things up.

35. Or rather destroyed. If you literally threw it away, the enemy might find it.

19 Automated Ciphers

Lewis Carroll's sliding Telegraph Cipher was a crude attempt to automate encryption, but a start had already been made on that job – in a rather more sophisticated way – several hundred years earlier. In the 15th century Leon Alberti (who we've met before in chapter 13) invented rather a neat little cipher wheel by putting together two copper discs of different sizes, each with the alphabet engraved around its rim. By rotating the inner disc so the alphabets were no longer aligned, he had himself a handy Caesar cipher machine that allowed him to read off a shifted alphabet at a glance.

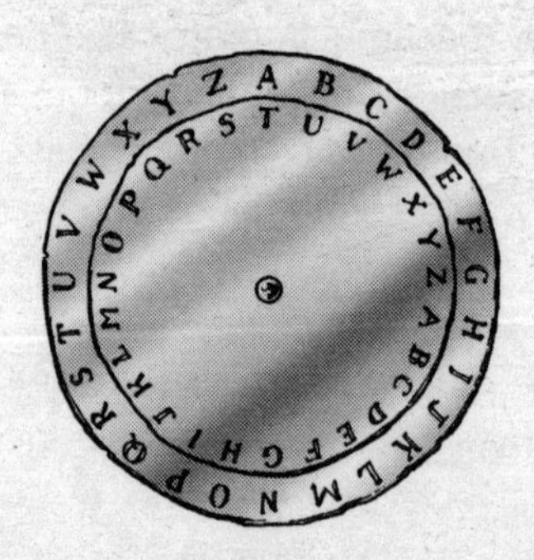

In fact, as he quickly discovered, he had something even better than that. Alberti came up with the idea of varying the setting on his wheel *during the course of an*

encryption. This effectively turned the wheel into an automated version of the Vigenère Square – and one far less prone to human error.

(I don't know if Alberti ever took the step of using keywords, but with a device like that he certainly could have. If his keyword was ITALY, he'd first set the A on his disc to coincide with the I of ITALY, then read off the first cipher letter of his secret message. Then he'd reset the A so it was opposite T etc. and continue with his encryption.)

Moving discs of this sort didn't generate ciphers that were any more secure than their standard counterparts, but they did lay a foundation for the development, some 500 years later, of a cipher machine that looked – for a time – as if it was just as unbeatable as the one-time pad cipher. It was invented just after the end of the First World War by a German named Arthur Scherbius. He called it *Enigma*.

Although it went through several versions, the finished Enigma machine looked much like a very clunky, very early, electronic typewriter. But it was a lot heavier than any typewriter I've ever used and at first glance it seemed to have two keyboards. Closer examination showed that the second 'keyboard' – on a panel just above the first – wasn't a keyboard at all, but an arrangement of alphabet letters that lit up individually from behind. Apart from this feature there was a peculiar mechanical dial and some neatly made housing for what looked like three cogged wheels.

The workings of Enigma were complicated. When you typed a particular letter on the keyboard, instead of punching up the same letter as a normal typewriter would do, Enigma sent an electrical pulse through three scrambler wheels, each one essentially a sophisticated version of Leon Alberti's cipher disc. These wheels were so designed that they could be removed from the machine and interchanged with one another in any sequence you wanted.

With all three wheels in place, each one changed the letter you'd just typed into another letter entirely, then automatically moved a place so that next time you typed a letter (even the same letter) it would be encrypted in an entirely different way. This set-up produced more than 17,000 different cipher set-ups for any given placement of the wheels, but Enigma didn't stop there.

When your letter travelled, so to speak, through the three scramblers, it hit an ingenious electrical reflector that sent it back again through the scramblers, *but by a different route*, before diverting it to the output board where the encrypted equivalent of your typed letter then lit up.

To make the cipher even more secure, Herr Scherbius added an electrical plugboard between the keyboard and the scrambler wheels. This board accommodated six cables, which you could plug in or out as you saw fit. Every time you did so, a particular letter typed – but not always the same letter – would be changed to something else, *even before it reached the scrambler wheels.*

There were a few other bits and pieces in the machine,

but these were the main elements. By putting them together, you created a cipher with the equivalent of *10,000 billion*[36] different keys.

What came out of a particular Enigma machine depended on the way you set it in the first place. Any other Enigma set in the same way could quickly decode the ciphertext for you. But a cipher-buster equipped only with paper and pencil would have a very hard time indeed. Arthur Scherbius thought his invention was unbeatable and sold it as such. The German military believed him and eventually bought more than 30,000 Enigma machines to help them revenge the defeat they'd suffered in the First World War. Settings for the machines were contained in a hefty codebook and changed every day.

Since the 1914-18 War ended, British cryptoanalysts had been routinely cracking open every German secret message sent and seemed set to go on happily decrypting them forever. But then, in 1926, they were suddenly faced with a mass of messages that beat every decryption technique thrown at them. The Germans had started to use Enigma. Seven years later, it was in the hands of the Nazis. But by then, incredibly, the cipher had been cracked.

Although Hollywood likes to claim the Americans did it and the British now largely believe Enigma was cracked by their people at the Government Code and Cipher School in Bletchley Park, the real credit lies with the Poles.

Poland had been monitoring Germany's secret messages just as keenly as the British and were twice as worried when those messages began to bewilder their cipher experts. They quickly realised the Germans were using Enigma machines and promptly bought one of their

36. Still the British billion – one million million.

own. But the only model they could get hold of was the commercial version, which was distinctly different to the one used by the German military. Examining it gave Poland the basic principles of Enigma, but absolutely no clue as to how the military version actually worked.

From 1926 to the end of 1931 there was no sign of a breakthrough. But then a French secret agent managed to buy some German documents that allowed the Allies to build a replica of a military Enigma. This was a start, but still a very long way from beating the German encryptions. Without the specific machine settings that acted as Enigma keys, there was no way of reading the cipher.

In fact, the Poles never did crack the cipher. Instead a 23-year-old genius named Marian Rejewski cracked the settings. Using some hideously complicated logical deductions, which I have absolutely no intention of trying to describe to you here, he managed to isolate the problem of the scrambler wheel settings from the problem of the plugboard settings. Taken together, there was no way of finding them. By taking them separately,

Rejewski was able to analyse patterns in hundreds of German messages that pointed to each. It took him a year to get his system right, but by the end of that time, he could work out new German day settings before the end of the day they were introduced.

With the settings and a duplicate Enigma machine, Rejewski could read German secrets as easily as if they'd been published in the London *Times*.

20 Computer Cryptography

By the time world war again broke out in 1939, the Nazis had introduced extra scramblers into their Enigma machines which complicated the ciphers so much that Rejewski could no longer extract the keys from them. But that was really just a matter of resources. The young Pole had shown Enigma wasn't perfect and that was enough to keep other experts trying.

One of them, a reserved Englishman named Alan Turing, constructed electrical circuits and networked several Enigma machines to help discover the keys the Nazis were using on their new, improved models. Another, the mathematician Max Newman, tackled an even more complex cipher machine the Nazis introduced later in the war. Using several of Turing's ideas, he designed a piece of electronic equipment that would not only crack the existing German cipher, but adapt itself to tackle different problems that might arise in the future.

An engineer named Tommy Flowers linked together 1,500 electronic valves to make Newman's vision a reality. He called the result Colossus and had it fully operational by the end of 1943. Nobody realised it at the time, but Colossus was the future of cryptography and the future of the world. Colossus was the first-ever electronic digital computer[37].

Colossus

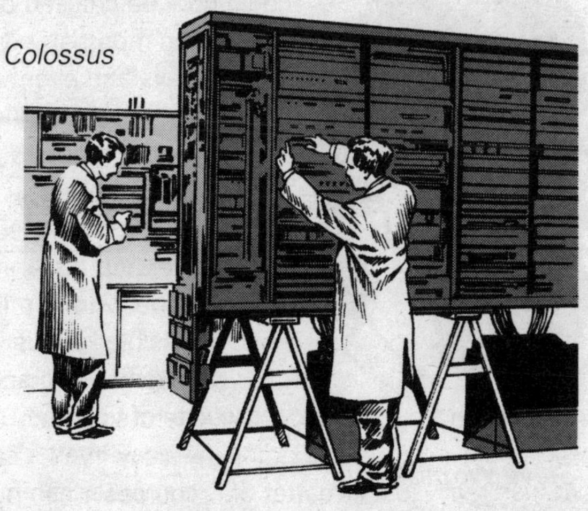

The arrival of computers made an immense difference to the arts of cipher-making as well as cipher-breaking. Computers work far faster than any mechanical cipher machine ever could and can be programmed to create ciphers of extreme complexity. This put the cipher-makers – i.e. the government and the military – way ahead of the game while they were the only ones who had computers. But then the first made-to-order commercial computers began to appear around 1951, which meant large companies could get into the act. By 1975, the first microcomputer was on sale and just a year later Steve Wozniak was soldering together the world's first personal computer in a Silicon Valley garage.

It was the spread of personal computers that made a nonsense of the old style of cipher. Systems that would

37. I know the Americans boast they built the first electronic computer in 1945, but they only get away with that claim because the British classified Colossus top secret, burned the only plans for its manufacture and have been far too polite to contradict their trans-Atlantic cousins ever since.

Steve Wozniak

once have taken centuries to break using paper and pencil, could now be cracked open within months, weeks, days and eventually hours or minutes using patient computers that would make trial and error substitutions at a faster and faster rate. A cipher with a million possible keys is well beyond the practical capacity of you and I, but today even a small computer can make mincemeat of it in no time.

Faced with this situation, cryptologists began increasingly to use computers to *make* their ciphers as well as break them. If a million keys wasn't enough, why not have your machine generate a cipher with a hundred million keys, or a thousand million keys, or a million million keys? Crank it up enough and even another computer will find it hard going to crack the ciphertext. Encryption takes up minimal computing power compared to decryption once you've worked out the algorithm for your cipher.

The computer giant IBM saw the possibility of profit in developing such an algorithm and selling encryption software to the business world. The package they came up with was called Lucifer – because it was devilishly clever: even suits can have a sense of humour – and in 1976 it was officially adopted as America's 'data encryption standard' the basic cipher used by US businesses to send sensitive information.

In theory, Lucifer could generate ciphertext that had just about as many possible keys as you wanted. But this panicked the country's National Security Agency, which had nightmares about ciphers even their enormous computers couldn't crack, so IBM was only allowed to develop Lucifer so far. The old devil ended up with 'only' 100,000,000,000,000,000 keys, enough to keep your business rivals at bay, but not enough, in the Land of the Free, to stop the government sticking its nose in.

But actually the government proved the least of business worries. Computers might have (more or less) solved the problem of large scale commercial encryption, but the age-old nuisance of key-distribution still loomed large. If you had Lucifer on your computer and your friend had Lucifer on his, you could send each other messages to your heart's content. But to make sense of them, you also had to send the key and the only really, really, really secure way to do that is to hand it over in person. Who has the time to do that? And if you have, it's then easier to whisper the message in his ear.

So key distribution remained the biggest hassle for cryptographers – governmental, military and commercial – right up to the time a bright spark named Whitfield Diffie figured how to get around it altogether.

The first step on Diffie's journey was an interesting bit of logic that showed two people don't have to have the same key to unlock a cipher. Suppose you put your secret message in a cash-box and lock it with a padlock (equivalent to your cipher). Only you have a key to the padlock, so you can send the boxed message quite safely to a friend. When it arrives, your friend can't open the padlock, because he doesn't have a key, but he can put a second padlock on the box and send it back.

Once again the message travels safely because only you have the key to Padlock 1, while only your friend has the key to Padlock 2.

When the box comes back, you take off Padlock 1 using your key and send the box off a second time. The message *still* travels safely because it's guarded by Padlock 2. But when it reaches your friend for the second time, he uses his key to open Padlock 2 and consequently reads the message with ease.

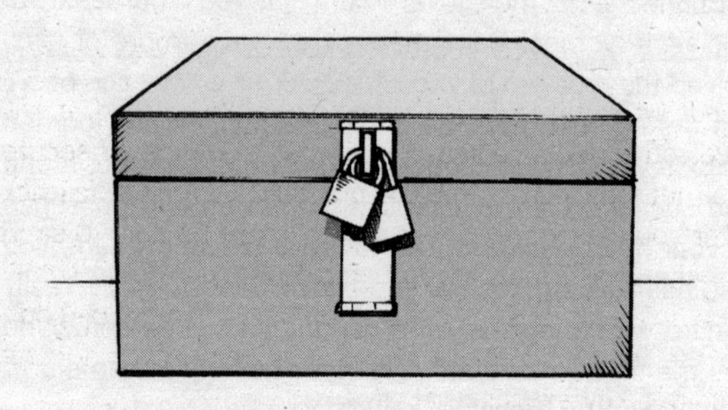

You can't *quite* use this method with ciphers – you might try working out why, but if you get bored with that, the answer is in the panel – but the basic logic got Diffie thinking and eventually, with the help of two colleagues named Hellman and Merkle, he used one-way maths functions to devise a method whereby a cipher key could be passed on with complete security (by phone for example) even if a third party overheard the conversation[38].

38. It was actually Hellman who made the breakthrough on this, but all three agreed to share the credit since the final step relied completely on the work they'd already done together.

Why Cipher Padlocks Don't Quite Work

While your friend can put on his padlock *beside* yours (which means you can remove your padlock without problems when the box comes back) the nature of ciphers means that he could only put a cipher padlock *on top of* yours. Since you don't have a key to his padlock, you can't get at yours to remove it.

But while this was an enormous step forward, it still involved an inconvenient key exchange, however secure the method. You can see the problem: you might fall sick after you've sent your message and not be able to send the key at all. Actually I'm trying to let you down lightly here. In a serious espionage situation, you might be shot. So the ideal situation is to have a cipher that doesn't require a key exchange at all.

Which is exactly what Diffie dreamed up when he hit on the idea of an *asymmetric* cipher. Up to this point, all ciphers were decrypted using exactly the same key that had been used to encrypt them in the first place. (You simply reversed the process.) But Diffie wondered what would happen if you created a cipher that was encrypted using one key and decrypted using another.

What would happen, as it turned out, was a free-for-all system now known as *public key encryption*. Basically you have two keys to your own personal computer-generated cipher. Both are sets of numbers. One set is your decryption key and you keep that strictly secret. The other is your encryption key and you publish that on the

Web for anyone to see. If I want to send you a secret message, I use the (public) encryption key to cipher it just like anybody else. But only you can read it, because you're the only one who has the (private) decryption key.

It was Diffie who came up with the basic idea, but three fellow Americans, Ronald Rivers, Adi Shamir and Leonard Adleman, figured out the maths that allowed computers to do the trick[39]. The cipher became known as RSA public key encryption. It uses a system based on factoring – the discovery of two prime numbers that you multiply together to make the key number. Where the key is very large, this is a nightmare even for computers. In fact, if the key is big enough, you could put every computer in the world on the job and still not have an answer in your lifetime. Unless, of course, somebody figures out a quick way of discovering prime factors.

39. Or maybe not. UK sources maintain three British cryptographers, James Ellis, Clifford Cocks and Malcolm Williamson, were the first to develop public key encryption, but their work was sensibly kept secret by the British authorities.

21 Pretty Good Privacy

Throughout the 1980s, the only people who owned computers big enough to use RSA public key encryption were the government, the military and a handful of the largest multi-nationals. But the Internet was quietly growing at this time and by the early 1990s, the general public were getting to hear about it.

And what they heard, they liked, because the Internet promised a new way of sending letters that was faster, cheaper and easier than anything offered by any post office on the planet – email.

Since it was first introduced in 1988, the use of email has expanded to a point where it has now become arguably the most widely used means of communication in the world. And therein lies a danger. Because while it's quite impractical for any government to stop and search every physical letter that passes through the postal system, it's perfectly possible to monitor every email[40]. Which means that if you regularly use email, your privacy is pretty well shot.

Somebody who worried about this was an American named Phil Zimmerman. After a while he worried so much that he set out to do something about it.

40. In fact, if you read my Spies Handbook (published by Faber & Faber) you'll find some governments are already doing it.

Zimmerman's idea was to make RSA public key cryptography available to everybody. That way you could send and receive emails nobody could snoop through.

There were several obstacles in the way of this goal, not least the fact that RSA encryption was a patented product, which meant he would have to get a licence from the owners, RSA Data Security Inc. But in the early days, Zimmerman had a more urgent worry. He needed to solve the one problem that prevented so many people using RSA – the need for heavy duty computing power. Even with the newest PCs, encrypting with RSA took more time than most people wanted to spend, especially if they had a lot of emails or a particularly long message.

He got around that one in a very ingenious way. He figured there wasn't much wrong with the usual way of enciphering a message. The really vulnerable area was the key which, with conventional cryptography, had to reach your recipient safely. So suppose you encrypted your message the old way (which could be done very quickly using a computer) and sent the key with it *but encrypted the key using RSA.*

Encrypting the message takes no time at all because you used old-style cipher methods. Encrypting the key takes no time at all, because it was short. But by using RSA, nobody can crack your key, which means your message is secure as well.

Zimmerman put his ideas together in a nifty software he called PGP, which stands for Pretty Good Privacy. Then he did something really wild. During the high summer of 1991, he asked a friend to post the package on an Internet bulletin board. Anybody who wanted it could download PGP for free.

For a while almost nobody noticed, but PGP was a time bomb waiting to explode. Within two years the roof fell in on Zimmerman. RSA Data Security Inc were threatening to sue for infringement of patent (since Phil hadn't actually got around to asking for a licence) and the FBI (Federal Bureau of Investigation) had launched an investigation into the possibility that he'd illegally exported a weapon[41].

It all sorted out eventually. In 1996, the Feds dropped their case and Zimmerman reached a settlement with RSA Data Security who issued him a licence. As a result, you can still get a copy of PGP on the Internet (from www.pgpi.org) free of charge so long as it's only for personal use. The download is around 5.3 MB and the application is a little gem.

Once you're equipped with PGP, just about everything you've learned from this book becomes obsolete since your computer will do it for you faster and better than you could ever do it yourself. (But knowing how to is useful for those times when you can't get to a computer and your contribution to my next royalty cheque is very much appreciated.) In fact, you have in your hot little hands almost as secure a method of encryption as is currently available anywhere on Earth.

Whether it will stay that way is an open question. The history of cryptography is littered with ciphers that were considered unbreakable in their day, but eventually collapsed under cipher-buster attack. A really major hike in computer speed might possibly pose a threat to public key encryption. A mathematical breakthrough in prime factoring certainly would.

But by the time any of that happens – if it ever happens at all – the vision of an Oxford University physicist named Artur Ekert may have become a commercial reality. In

41. The US Government very wisely included certain software within its definition of munitions and insisted you needed a licence before sending it out of the country.

1990, Prof Ekert suggested that quantum entanglement – the peculiar behaviour of certain sub-atomic particles – might be used as the basis of a totally unbreakable cipher system.

With the Ekert system, the key is sent along with the message as a stream of randomised particles. Any interference with the stream before it reaches its intended destination changes the key, rendering it unusable by the cipher-buster. There is mathematical proof that such a system must remain absolutely unbreakable no matter what scientific or technological advances might be used against it. But the question that taunted cryptographers since 1990 was whether the Ekert theory could ever really be put into practice.

Then in April 2004, the sum of 3,000 euros was securely transferred between Vienna City Hall and the Bank Austria Creditanstalt using quantum key distribution. The age of the absolutely, totally, guaranteed unbreakable cipher has at last arrived. Although, as you may discover in the final chapter of this book, it could actually have arrived eight centuries ago.

22 The Voynich Manuscript (And How You Can Become a Celebrity)

If you're not fairly expert in ciphers by now, you haven't been paying attention. But assuming you are and were, the time has come for you to use your knowledge, insights and intuition to earn yourself some fame courtesy of the Voynich Manuscript. So what's the Voynich Manuscript?

Back in 1912, an American collector named Wilfrid Voynich was rooting around a hoard of illuminated manuscripts he'd discovered in an ancient villa near Rome when he discovered a bulky (102-page) package of pages written entirely in cipher and illustrated by numerous drawings.

The style convinced him the work dated to the 13th century and the drawings suggested it might be a work on natural philosophy. Voynich promptly bought it.

A history of the manuscript gradually emerged. It's now known that King Rudolph II of Bohemia – a collector of dwarves and giants – bought it in 1586, apparently under the impression that it was the work of the 13th-century English monk, philosopher and astronomer Roger Bacon. Sometime after 1608, it passed into the hands of Jacobus de Tepenecz, the director of Rudolph's botanical gardens. From there it went through several owners until it eventually ended up in the care of the Jesuit College at the Villa Mondragone, where Voynich eventually found it.

The original manuscript comprised one hundred and sixteen 15 cm x 23 cm vellum pages of which 102 remain. It contains drawings of unidentified plants, various (possibly astronomical) charts, tiny naked people in what look like rubbish bins and a design for the most bizarre system of plumbing you're ever likely to see. But it was the encrypted text that really intrigued Voynich, who handed copies to several experts in the hope that one of them would break the cipher.

It looked like an easy enough job. But the cipher-busters quickly discovered the original plaintext had not been written in Latin, as most of them confidently expected, nor Old English, Medieval French, High German or, indeed, any other known European language. This meant – as you've already discovered in this book – that standard decryption techniques could not be applied.

Various suggestions were made, but none resulted in a satisfactory decryption. The original manuscript was bought in 1961 by a New York collector who subsequently donated it to Yale University. To date, there's still nobody in the world who knows what it says.

Because if you're the one to crack the Voynich, you're guaranteed instant fame and can write your own ticket as a codebreaker thereafter. So give it a try – it's fun.

To help you along, here are some suggestions that have *already* been made:

- The manuscript is a book of alchemy.
- It's a book on herbs.
- It's a rare book by the Cathars (a persecuted religious sect in medieval Europe).

- It's just a meaningless collection of characters created by a scoundrel to make money.
- It's nonsense, written by a medieval quack (fake doctor) to impress his patients.
- It's not in cipher at all – we just don't recognize the language.
- It was coded before it was ciphered.
- It came from another planet.
- The original language is just made up.

And here are a few facts about the manuscript that may be helpful:

- Computer analysis suggests the plaintext contained not one language, but two (or at least two very different dialects of the same language).
- The text has no apparent corrections.
- Most of the plants illustrated have never been identified.
- Some of the symbols in the manuscript are similar to alchemical symbols (alchemy was the forerunner of modern chemistry), Latin shorthand or early Arabic numerals.
- The manuscript appears to be divided into sections loosely categorized as Astronomical, Biological, Cosmological, Herbal and Pharmaceutical.

Here are some sample pages of the manuscript for you to work on. You can find others at www.voynich.nu.

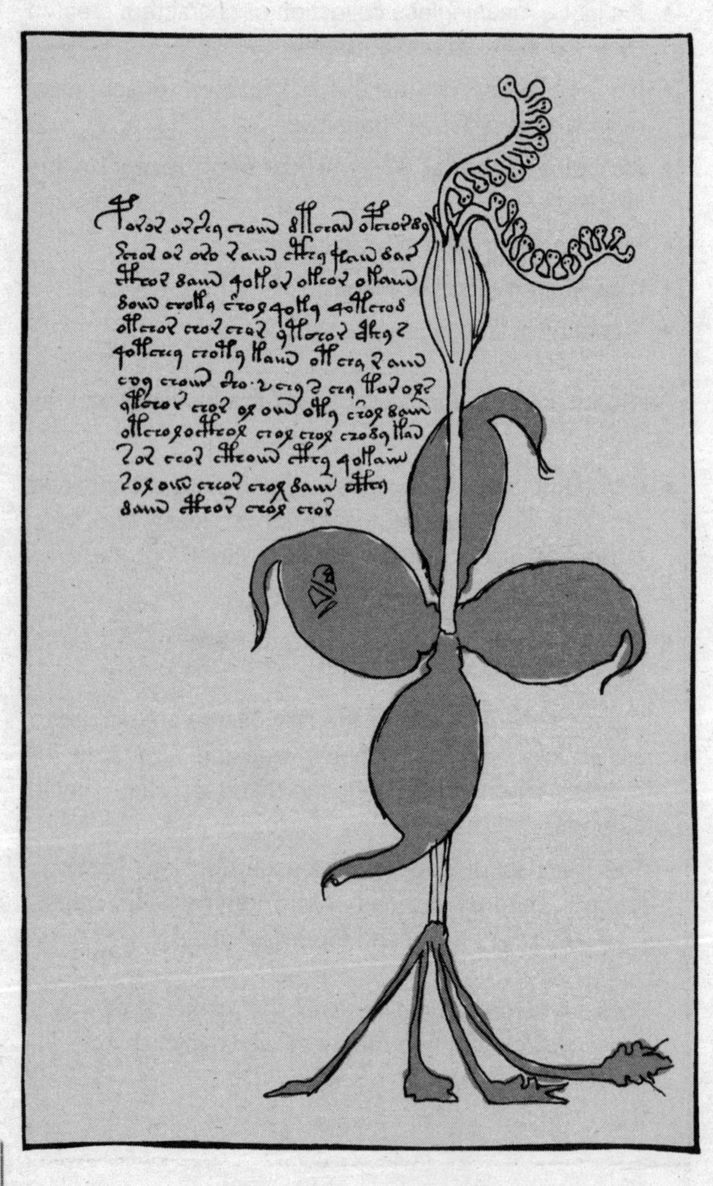

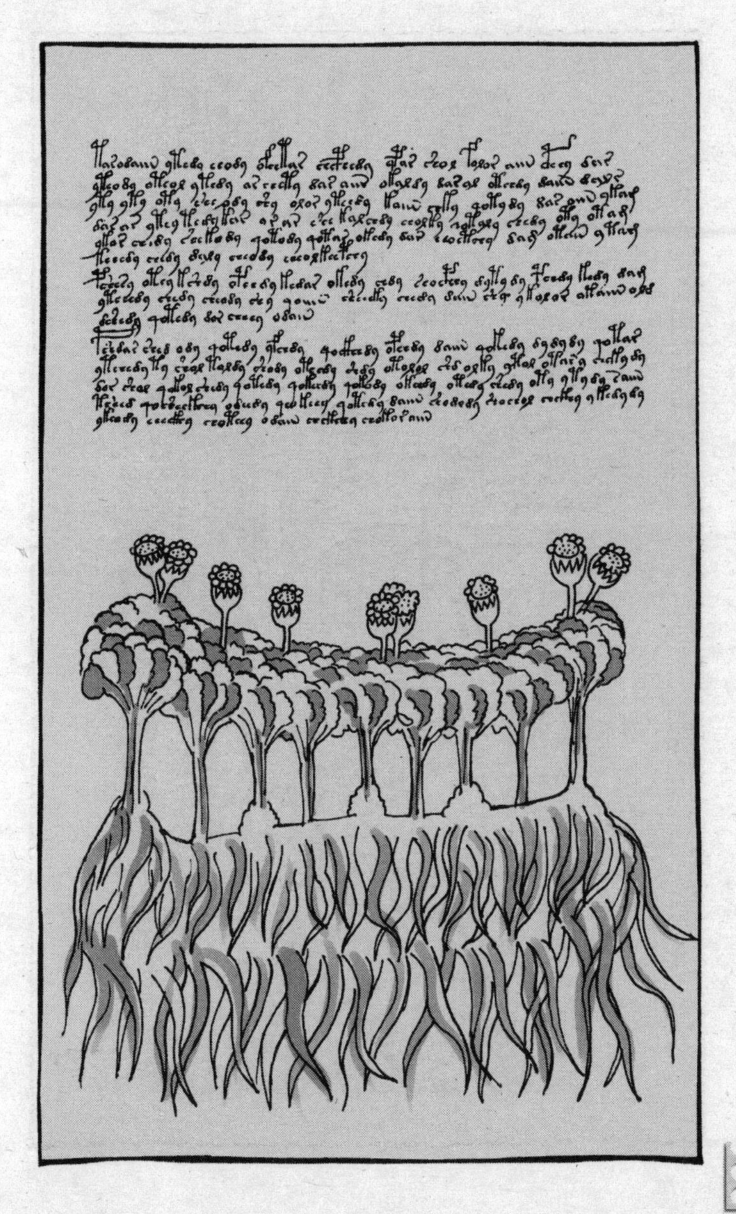

Here's a close-up of the elaborate text to help you crack it.

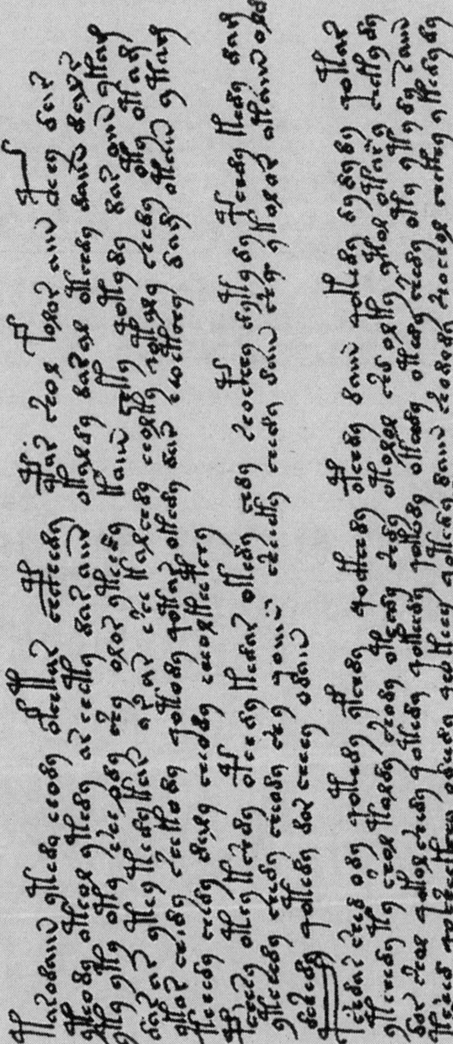

Afterword

Confession time. I didn't make the Voynich manuscript the basis of our final chapter just to give you some decryption practice. I included it because it gives me a warm feeling every time I think about it.

Because while the exact dates are uncertain, you can be sure the Voynich was encrypted long before the difference engine was a glint in Babbage's eye.

I like that. I like even more the fact we haven't cracked it yet. It sort of gives me hope.

Impressed though I am by PGP and the prospect of quantumly entangled ciphers, all that is computer business. And when you study the history of encryption, one thing stands out.

It's more fun when you do it yourself.

Index